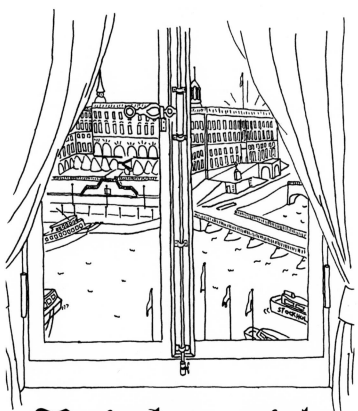

Reindeer with King Gustaf

What to Expect When Your Spouse Wins the Nobel Prize

ANITA LAUGHLIN

Wyatt-MacKenzie Publishing, Inc.
DEADWOOD, OREGON

Reindeer with King Gustaf
What to Expect When Your Spouse Wins the Nobel Prize
by Anita Laughlin

Illustrations by Robert Laughlin
Cover painting by Diana Bradley
Digital file preparation of cover painting by Susanne Weihl
Family photo by Margaret Martin

Wyatt-MacKenzie Publishing, Inc.
DEADWOOD, OREGON

www.WyMacPublishing.com
(541) 964-3314

Requests for permission or further information should be addressed to:
Wyatt-MacKenzie Publishing, 15115 Highway 36,
Deadwood, Oregon 97430

Acknowledgments

The idea for this book originated during a dinner conversation among Nobel laureates and spouses, which took place several years ago on Lake Constance, in Lindau, Germany. It was at this international meeting sponsored by the late Sonja Countess Bernadotte of Wisborg, that our memories of the Nobel experience were openly shared with tenderness and hilarity. Thankfully, Bettina Countess Bernadotte of Wisborg continues the Lindau tradition today.

My gratitude extends to the hundreds of people associated with the Nobel Foundation along with the chefs, waiters, florists, entertainers, musicians and Swedish students who pool their talents to create the magic of the Nobel week's events.

Thanks also goes to those who kindly gave me early reviews of my book: Doug Osheroff, William Phillips, author Mary Reath, publisher Thea Selby, Bob Bass, along with my many family members who continue to generate spirited Nobel stories—some eleven years later!

I am also indebted to artist Diana Bradley for the cover painting; to Susanne Weihl who assisted with the digitizing of the cover painting; to my copy-editors Carolyn Perry and Bobby Maltese. Special thanks goes to my good friend and editor, Lisa Pliscou, for her encouragement and sage insights. I am also grateful to my publisher, Nancy Cleary, who has been so supportive of this project.

For their love and support, I thank my parents, two sons, and of course my husband Bob, who have brought such joy to my life.

— Anita Laughlin
Stanford, California, February 28, 2009

Illustrations by Robert Laughlin

Contents

FOREWORD

OCTOBER 2008

I am writing to give my impressions of the thoroughly amusing book Anita Laughlin has written of her experiences as the spouse of a new Nobel laureate. Perhaps part of the reason I found her manuscript to be so captivating is that I had received my own Nobel Prize just two years before Bob Laughlin received his prize, and reading how Anita described events that I had lived through was both amusing and also thought provoking.

For a brief moment I thought that perhaps I should write my own manuscript, if only to preserve my memories of unique events. However, Anita has done it so well, my own recollections would pale in comparison. I, for example, have no idea what food was served at any of the banquets held during Nobel Week. I also did not keep a log of the various events and problems that had arisen (but they certainly did arise). I only remember whom I sat next to at one of the banquets, and that is because she is the Queen of Sweden.

I also enjoyed Anita's tale (which of course began with a call from Stockholm at 2:30am) because her perspective on the events surrounding the awarding of the Nobel Prize was quite different from my own. I never once worried about how my hair looked, and the only clothing I purchased for the occasion was my white tie and tails.

I think we both agree that the banquet that was the most fun was the one hosted by the students, although Anita could not have appreciated how wonderful it was to be surrounded by pretty young Swedish students. Anita's description of the induction of the new Nobel laureates into the Order of the Ever Smiling Jumping Green Frog was remarkably accurate, although I don't know if Bob told her how really tiring it was to hop around on the floor like a frog for five minutes. I was concerned for some of the older laureates.

The Nobel Prize makes one an instant celebrity. Yet very few people have any knowledge of the series of events surrounding the granting of this high honor. I think Anita's description of these events is both accurate and entertaining, and should prove fascinating to almost anyone, even if they didn't get all A's in their science classes.

— Professor Douglas D. Osheroff
STANFORD UNIVERSITY

CHAPTER 1
Receiving the Call

Confident as I am googling for San Francisco street directions or for carrot cake recipes, there was one eventuality for which googling did little to assist: *What happens when your spouse wins a Nobel Prize?* This memoir is an attempt to demystify this occasion, as there is no website that can do it justice. There is no official handbook. Your loyal friends cannot give you much guidance. You are on your own.

When you realize that your spouse is one out of six billion to receive this recognition, and you are one out of six billion who by default will be on a plane to Stockholm, it is a humbling feeling. I am here to tell you that there is no way to predict how the experience will change your life individually or as a couple. The prize itself is grand, but perhaps the journey to receive it is as entertaining and important.

On December 10, 1998, my husband, Robert Betts Laughlin, walked over to King Gustaf of Sweden, bowed his head to receive his Nobel Prize medal, accepted his leather-bound diploma, shook hands, and then turned to bow in appreciation to the members of the Swedish Academy, other Nobel laureates, and to an audience of millions watching the ceremony on television from all over the world. I sat with my family in the second row of Stockholm's grand City Hall, tearfully acknowledging Bob's steady gaze in my direction as he bowed onstage. He had worked so tenaciously to earn the right to stand among the recipients, I could almost detect his sigh of relief and pride that his creative work had finally been recognized in this way. His was a Nobel Prize in Physics for work he had done as a young theoretical physicist.

From 1978 to 1980, Bob worked as a postdoctoral fellow at Bell Laboratories in New Jersey. At the time, Bell Labs was a prestigious lab. It was a cleverly designed facility with comfortable common rooms in which famous scientists who had been recruited from all over the world could informally meet to discuss their ideas. The company most analogous to Bell Labs today is the sprawling Google complex of pampered techies located in sunny Mountain View, California.

When he arrived at Bell Labs in 1978, Bob worked inside the theory group and could choose his own projects. John Joannopoulos, his thesis advisor from graduate school days at MIT, had strongly advised Bob to work with Bell Labs experimentalist Mark Cardillo, whose work John thought was particularly good. Cardillo had a new technique for exploiting the quantum nature of atoms specifically, using their wavelike properties to study surfaces. So my husband began work in this area, attended seminars, and politely appeared for tea and cookies once a week with his colleagues to chat.

It was at one of those civilized tea parties that Bob talked to another Bell Labs colleague, Dan Tsui about a paper that Klaus von Kiltzing had written. Klaus von Klitzing was an experimental physicist on the faculty of Würzburg University in Germany. Dan Tsui was quite excited about this particular paper and encouraged Bob to read it. Bob had followed Tsui's work over the years and greatly respected his opinions, so he got to work. Two weeks later, Bob gave a Bell Labs Journal Club presentation which was a sort of a show and tell opportunity for colleagues to hear a formal lecture about research going on at the lab. Bob showed his recent work on the "gauge argument" which was a simple math theorem showing that "von Klitzing's accurate number was equivalent to the charge of the electron." It should have been impossible to get a third of von Klitzing's number because the charge of the electron is indivisible.

Bob's job search in the winter of 1980, combined with the sudden death of his father, made it a dismal period for us. I had found a job at a private school for disturbed children and had my hands full with a class of eight seven-year-olds who were diagnosed as psychotic or schizophrenic. One of them believed himself to be a T-rex dinosaur. Another would take on the personality of her murderous mother whenever she felt afraid in class. A third child had climbed up to the roof of the school and threatened to jump off while dangling from a drainpipe. I was hanging by a thread myself, professionally speaking. Bob and I longed for a change of scenery and Bob in particular missed weekend hikes in the Sierra Mountains. Luckily, Bob easily found employment six months later at Lawrence Livermore Labs in California.

When he arrived at Livermore, however, he was required to sit out in a stuffy portable in the parking lot for six months until he received a security clearance to enter the facility as an employee. This banishment proved to be a blessing.

While an outcast on his own portable island, Bob received an early pre-publication look at an article from Bell colleagues Dan Tsui and Horst Stormer. In it, they described the first observation of the *fractional quantum hall effect* noting: "You might be interested…" They had understood that: "The gauge argument precluded such an effect unless the electrons were forming a new state of matter, hence their impish challenge." Bob was encouraged by the Livermore management to work on understanding this new effect in addition to doing his weapons-related research. The significance of this finding is still impossible for me to understand but Bob has tried to simplify it for me this way: "The effect Tsui and Stormer discovered involved magnetism. The new effect looked similar to Klaus von Klitzing's *quantum hall effect* (for which he would win a Nobel Prize in 1985) but had numbers that came out *one-third* of von Klitzing's measurements. The Tsui-Stormer effect occurred at high field strengths at which the von Klitzing effect was impossible. This meant that the electrons were organizing themselves quantum mechanically in a way never before witnessed and unknown to science."

Over the next two months Bob thought about these results—much of the time in a trancelike state beneath our golden delicious apple tree in the backyard. The house literally could have burned down and Bob would not have budged. He quickly wrote two papers on the subject, the second of which was a mathematical description of this self-organization using a single equation. This phenomenon acquired the nickname "1/3 effect" and later became well-known as the *fractional quantum hall effect*. A single equation about anything so complex is profoundly impressive.

Bob's equation was very simple, but the underlying concepts involved quantum mechanics, which is not very simple. In quantum mechanics, you cannot say where electrons are, but only the probability that they should be in certain places. Bob's equation describes those probabilities very succinctly. The equation involves only fourteen symbols, but describes 100 million electrons! Bob elaborates: "The key implication of this description is that these electrons can move together collectively in such a way as to masquerade as a particle carrying one third of the electron charge. This is shocking because everyone thought that the electron charge was indivisible. This experiment and the theory that goes with it thus show that this premise of fundamental physics is not true. This in turn has potential implications for fundamental physics, as the quark, the elementary constituent of protons, carries one third of the electron charge. The suggestion here is that the quark may not be fundamental after all."

I confess to ignorance about quarks but have taken his word for it that they exist. Not ever having taken a physics course, I am indebted to Bob for the many hours he has spent trying to bring me up to speed on his discovery. Some of the dinner discussions in our house over the years have been simply meteoric in scope: whether we are discussing physics, Middle Eastern policies, American politics, the philosophy of Marcus Aurelius, information age glitches, or murder mystery plots, Bob always voices his informed opinions with lightning-fast clarity. There is nothing he has not thought about. Through him, I have actually acquired some number sense. For example, the synchronicity of numbers now fascinates me. Bob and I had our first date on October 13, 1977. He received his phone call from Sweden on October 13, 1998. The prize was for his "1/3 effect" which he shared with Tsui and Stormer, as one of three recipients.

Bob was on the short list to receive a Nobel Prize ever since his groundbreaking theory emerged in 1982. When he finally did win, he was forty-seven years old. We had been married twenty years, had two spunky sons, and were living in a light-filled home on the Stanford University campus surrounded by gardens of roses and dahlias and a dozen redwoods.

When writing any memoir, there is always the "x-y chromosome effect" and I need to be clear that my journaling of the months following the announcement are entirely from my point of view, and may or may not reflect what others have experienced. I have often been asked by colleagues, families of students, friends, and complete strangers both in the United States and in other countries about my version of what happened. Now on the eleventh anniversary of Bob's prize, I consider it fitting to chronicle my personal reflections of our long, loving relationship and how we responded to this extraordinary event in our lives.

Robert Betts Laughlin was born in Visalia, California, in 1950. Visalia is located in the hot farming region of central California. Bob's extended relations are musicians, teachers, lawyers, and doctors. His own father, David, was an attorney who, incidentally, read *Webster's Dictionary* every night before going to bed. He was a man who enjoyed vociferously debating issues at the dinner table with his children, Bob, Margaret, John, and Julie. He installed four blackboards in the hallway near the children's bedrooms to give them their own space to experiment with their ideas. David would occasionally work out math problems, inviting his children to join in and challenge him. Bob filled his blackboard with equations and sketches. David's engaging personality both modeled and validated the importance of intellectual inquiry for his four children. Probably as a testament to his influence, the Laughlin children later trained in law, medicine, computer science, electrical engineering, and theoretical physics.

Bob's mother, Peggy, worked in Visalia as a teacher and became an admired math mentor for other teachers. She, too, instilled a love of math, music, and inquiry in her children. Early on in Bob's life, Peggy became alarmed by his independent behaviors, which included running away at age three to sit on a tractor at a local dairy farm and jumping off the roof of his house at age four in his homemade Superman costume. At age five, he repeatedly asked his aunt to read him a book entitled *The First Book of Electricity* and later managed to secretly wire the power for his train set into the main electrical unit of the house. Bob read the *World Book Encyclopedia* on a daily basis (which he continues to do every night as an adult) and grew interested in a wide variety of things including pulley systems. His mother remembers once looking casually out the window and being startled to see hunks of metal and wood being quietly raised up to the second floor via a string pulley system. The destination for those items was the secret attic room in which six-year-old Bob tinkered.

Once Bob entered school, he was like a square peg in a round hole and did not welcome instruction from his teachers. Independently driven, he seemed to have his own learning agenda that was always two steps ahead of his peers and three steps ahead of most adults. As the number of his interests and questions increased, Peggy worried that Bob needed a more accelerated curriculum. In third grade, he refused to memorize multiplication tables and resisted learning standard algorithms for addition and subtraction. To this day, I am baffled by the way he balances a checkbook. In any case, Peggy and other local parents took action, setting up a private elementary school with a more rigorous curriculum.

As a teenager, Bob taught himself calculus and experimented with household chemicals out in the garage a practice that led to a hospital visit for burns to his hands. Recalling those days, Bob once asked a group of physicists in Helsinki, Finland, if any of them had ever created explosives as kids, and they all raised their hands. Some, like colleague Horst Stormer, had even lost fingers from rocketry mishaps.

Following a long family tradition, Bob attended the University of California at Berkeley. His father had urged him to major in engineering, but with characteristically tenacious independence, Bob followed a passion for theoretical physics instead. He attended Berkeley during the Vietnam era in which campus anti-war protests and explosions were an everyday occurrence. Ironically, during the mayhem of anti-war sentiment he received his draft notice and after graduating, enlisted into the army, where he was eventually posted in an office to type and answer the phone. It was while he was stationed in Schwäbisch Gmünd, Germany, that he taught himself German by listening to the radio. At the end of his tour of duty, Bob located some saltpeter and sugar, poured the mixture into his army boots, and ceremoniously blew them up. This, too, was a family tradition. His uncle, Dr. Hiram Betts, had been a surgeon in the Korean War and had obliterated his boots upon being discharged.

Shortly before Bob's own discharge, he signed up to take the Graduate Record Exam in Stuttgart, Germany. He had shared a locker in his barrack with three others, that they affectionately named the "munchies cabinet." They had stuffed a variety of contraband items inside such as cereal and candy, along with silverware, steel pots, speaker wire, sleeping bags, and ammo belts, which no doubt caused the cabinet to acquire a midriff bulge. The day before the test, there was an inspection and Bob claimed responsibility for the locker while standing remorsefully in a pile of cascading debris. Upon failing the inspection, Bob was given extra duty, which included hours of kitchen work. The major on duty repeatedly told Bob he would not be allowed to take any GRE test. But the next day the major relented and Bob went to take his exam feeling stressed and exhausted. He did poorly on the first half of the test, which might account for the graduate school rejections he began to receive in the mail. MIT finally decided to give him a chance and accepted him, but refused him financial assistance. It was at MIT several years later that Bob and I first met.

In contrast to Bob's home life in Visalia, my family crisscrossed the United States several times after I was born in Palo Alto, California. My mother had graduated from Stanford and my father was finishing his doctoral program at Harvard while I was a toddler in Cambridge, Massachusetts. He subsequently accepted college administration and publishing jobs in Oregon, New York, California, and Massachusetts, in addition to a World Bank posting as an educational consultant in Lesotho, Africa. In reaction to all of these moves, perhaps, I began to keep journals at a young age and wrote poetry day and night in the enclave of my bedroom, much to the consternation of my family.

Like Bob, I was fiercely private and independent, creative and internally driven. My adolescent emotions found their way into my poetry while under the influence of authors like Sylvia Plath, Emily and Charlotte Brontë, Flannery O'Connor, and Eudora Welty. And as with Bob's family, our dinner debates with my three siblings were commonplace. I also especially valued my own early exposure to the language of Shakespeare as I listened to vinyl recordings of plays and sonnets and attended live theater as often as possible. I graduated from Palo Alto High School with the school's top award for creative writing, the Ivan Linder Prize in recognition of my poetry and playwriting.

I attended Connecticut College, studied with poet laureate William Meredith, and spent hours listening to Ralph Vaughan Williams' music while reading John Milton's *Paradise Lost*. After graduating, my freshman roommate and I relocated to Cambridge, Massachusetts, where I enrolled at Lesley College to receive two masters degrees in regular and special education. My roommate, Susan Froshauer, worked as a lab researcher at MIT studying slime molds, which required her to be at work at night. We would meet to swim in the MIT pool at the same hour, which happened to be when Bob would come to swim. I have always felt that it was Susan's slime mold that actually led me to Bob.

Bob and I both swam in the fast lane, so my immediate assumptions were that not only was he a young, brilliant MIT student, he was probably a native Californian swimmer. Physically strong and aggressive in the water, in retrospect Bob appears to have been no less than a Michael

Phelps anomaly in the otherwise sedate MIT pool. I later found out that his mother had required him to be on the swim team and play water polo in high school as a way to mingle with other students.

Intrigued, I did a little mingling of my own and eventually introduced myself after Bob remarked about a guy swimming in the lane next to us in a leopard swimsuit: "He may look tough, but he keeps his suit on in the shower." I invited Bob out for Chinese food, and he accepted; shortly thereafter, I spent an entire month's budget on cooking him our first homemade dinner, and seven months later in 1978, we were married in the Stanford Chapel.

In order to be married in the chapel, we were required to take a psychological compatibility test (this was, after all, Stanford University). On this test (in separate rooms), we were asked to fill out "How would your fiancé respond to x situation?" and fill in a dotted rating of one to five. We filled out this test for each other's responses, and one for our own reactions. The minister then graphed the responses and delivered the bad news: "You two should reconsider this marriage. You are not the least bit psychologically compatible." I relayed this pronouncement to my mother that we had flunked our compatibility test, but within forty-five minutes, the wedding was still on. She had phoned the minister from three thousand miles away in New England. I have no idea what she said, but Bob and I have been married now for thirty-one years.

If I were asked what married couple in history Bob and I most resemble, after a slight pause, I would say John Adams and his wife Abigail Smith Adams. Adams was a lawyer, statesman, and gentleman farmer in the mid to late 1700s. In 1774, New Jersey delegate Richard Stockton referred to John Adams as an "atlas of independence." I had actual goose bumps as I read David McCullough's initial description of Adams' personality since it so closely resembles my own husband's finer points:

> "John Adams was also, as many could attest, a great-hearted, persevering man of uncommon ability and force. He had a brilliant mind. He was honest and everyone knew it. Emphatically independent by nature, hardworking, frugal… he could be high-spirited and affectionate, vain, cranky, impetuous, self-absorbed, and fiercely stubborn; passionate, quick to anger, and all-forgiving; generous and entertaining. He was blessed with great courage and good humor, yet subject to spells of despair, and especially when separated from his family."[1]

This description fits Bob to a tee. And like Adams, my husband is ambitious, can recite poets like Longfellow from memory, reads classical literature, speaks several languages, is a constant traveler and an orator, and is an ardent proponent of the rule of law in all civilized societies.

[1] David McCullough, *John Adams* (New York: Touchstone, 2001) 18-19.

John Adams' wife Abigail kept herself apprised of the course of events shaping the birthing of our country, offering her opinions (whether sought after or not) through daily letter writing to John over the course of many years. She was his rudder, his calm in the storm. Self-sufficient and with four children to raise, Abigail remained patiently at the family farm, "holding down the fort," while John traveled, fiercely debating his ideas and solidifying his political future. Abigail eventually left the farm for an extended visit with John in France. Thrust into the company of royalty, she was way out of her comfort zone adjusting to the protocols and conventions of the French court. I had similar feelings when I visited Sweden and found myself dining with royalty.

Much like John and Abigail Adams, Bob and I are temperamentally polar opposites, yet we are both intellectually curious and frequently opinionated. We also share a profound respect for one another's need for independent, creative growth. This is a feature of our marriage that was not "tested for" at the Stanford chapel. It is at the core of our relationship and has evolved over time to supersede any hurdles we have encountered as a couple.

Twenty years after our first date, on October 12, 1998, the director of the Stanford News Service called Bob at his office to inform him that he was on an "even shorter list" of possible recipients to win a Nobel Prize, and that Bob was to call him immediately if he heard anything.

Bob promptly threw the phone number away, believing that Stanford could not possibly win a third year in a row (Steven Chu 1997; Doug Osheroff 1996).

But early in the morning of October 13, the phone rang repeatedly in our son Todd's room. As a hip thirteen-year-old, Todd had a large, trendy Mickey Mouse phone in his bedroom, which, I will remark, is a good deal larger and more colorful than the iPhone he now carries. Todd shuffled sleepily into our dark bedroom, jiggled Bob's toes, and said: "Dad, there's some guy on the phone from Sweden. Can I go back to bed now?" Bob sat bolt upright and groped for the phone in the dark. We had turned off the ringer the year our sons both hit adolescence. I heard Bob reply to the caller slowly and calmly. "This is the best kind of news," he said quietly. Bob then conversed with two Swedish colleagues, Mats Johnson and Stig Hagstrom. They were on the physics committee in Stockholm and were reassuring Bob that this was not a prank. It was the real thing. Todd stood motionless. I began to sob as Bob hung up the phone. We embraced and Bob leapt up to tell our other son, fifteen-year-old Nathaniel, who had slept through the whole incident. Bob searched the waste-baskets for the News Service's business card. Todd asked me why I was crying and then announced that he was going back to bed. I found out the next morning that he had actually allowed the phone to ring "a long time, maybe half an hour. I wanted to be asleep!" His Mickey Mouse phone was later photographed when he was asked a half dozen times to retell his story. The Nobel archives expressed an interest in having the phone, but Todd wanted to retain his "piece of history."

Bob decided to wake up his department chairman, Blas Cabrera, and asked him to call the News Service. This was at around 2:50am Bob awoke both his mother, Peggy, and my mother, Carolyn Perry, with the same question: "Want to go to Stockholm?" The screaming at the other end was audible across the room, and at least in my mother's case, went on for twenty minutes. My father thought she'd won the lottery. I turned on the porch lights just as the phone began to ring off the hook. Bob was now very awake and in various stages of undress, talking, and shaving. I searched hard to find him a tie, clean shirt, and blazer. As these items were the extent of his "wardrobe," he dressed quickly. I went down to make him some French roast coffee and toast. He was unable to walk across the room without hearing the phone ring.

The prize had been announced at 6:00am eastern time, and 3:00am west coast time. At 3:30am, an intrepid Jack Hubbard from the Stanford News Service knocked on the front door. With remarkable trust, because I was still in my bathrobe, I let him in. He was immaculately dressed. He asked where the phone was, tossed his own cell phone on the desk, and explained: "I am here to get you through the day. I am here to answer the phone, set up newspaper interviews, be Bob's bodyguard and chauffeur. We need to keep him calm and not let things get overwhelming. I take my cues from you on this." Great, I thought. Calm cues. Right.

I ran frantically upstairs and threw on a generic denim tent dress that was the best I could do at 4:00am. The boys were resolutely asleep. I peeked into Nat's room and said: "Nat, your dad's won the Nobel Prize!"

"Oh, yeah, I knew that. Good night."

Within a minute, Jack's assistant appeared. He was a jovial older man who was also helping with phones and prioritizing media interviews.

By 4:30am, an hour in which I have never actually been awake, I threw some cinnamon buns in the oven and began to scramble eggs as people kept arriving, all congregating in the kitchen. They began taking pictures all over the house. One photographer was taking pictures in the kitchen. "You want pictures of scrambled eggs?" I inquired.

"Oh, certainly," she responded as my eggs then appeared on her laptop screen. "I push this button," she said, "and your eggs get digitally transferred to newspapers all over Europe. Everyone in the world now knows you're eating eggs for breakfast."

"Great," I choked and ran upstairs looking for another dress.

By 5:00am there were white television vans pulling up in front of the house. Two huge broadcasting pipes wrapped with cables were laid out on the front lawn along with several white dishes. People began pouring into the living room from all directions (reminding me vaguely of that house invasion scene in the movie *E.T.*) carrying tall lights, miles of cables, cameras, and microphones. Within fifteen minutes, approximately thirty strangers were taking up stations inside and outside the house with commanding authority. The phone rang every sixty seconds. A female reporter said she needed two things: coffee with milk and a bathroom to do her makeup. I sent her on her way and wondered if either of our sons was about to get a pleasant surprise. Returning quickly, the reporter sat on a corner of the couch, plugged in her earphone, and waited for her station's cue to begin interviewing Bob. Bob insisted he needed another tie and we both bolted upstairs to find one.

Miraculously, our sons were still asleep. Emotionally and physically, Bob and I were already exhausted. Frantic and elated simultaneously, Bob seemed to go on automatic pilot, answering questions with candor and ease when the television interviews began at 6:15am. Watching him, I felt tremendously proud of this brilliant and relatively shy man of deep courage and conviction. I knew him intimately as a person, but less intimately as a scientist. I was finding out new things about him. I was witness to a kind of evolution in our relationship that morning as I listened to him speak. I thought at the time that perhaps retaining a private aspect of oneself is fundamental to a loving relationship. I thought how fortunate we both were to have maintained our individuality, our creativity, and our private purpose, even though married. We had allowed one another to grow internally in secret ways the other could only imagine. It was a joyful, personal realization in that house full of whispering people, bright lights, cables, monitors, and cameras.

Jolted back to the immediate situation, I worried that we were running out of food for the mobs of television crews who kept tripping over my feet, and I went on a bagel run. Upon returning, I got the boys out of bed and I stammered in passing: "No holey jeans!" Todd, then thirteen, was reasonably compliant in a crisis and sat calmly with me in the kitchen watching his father on television as he was getting interviewed in the living room—a surreal experience to be sure. He decided this was all too exciting to miss by going to school, but his brother, Nat, announced he had a chemistry test to take, so off he went.

We were delighted to have Sandy Fetter, senior scientist and chair of the Stanford physics department, drop by from three houses down the street. Our good friend and astrophysicist Bob Wagoner also joined us. Outside, our neighborhood emergency coordinator was combing the street in his bathrobe with his walkie-talkie charged, wondering who had been murdered. A third friend of ours, Peggy Esber, arrived with the urgency of a Red Cross volunteer, generously bringing a gorgeous pink-and-white floral bouquet, French champagne, and more bagels.

I had phoned in the report of my absence from my

teaching job and awoke my favorite substitute, Jackie Andrew (also married to a physicist) who graciously accepted a pre-dawn assignment, and I asked her to try to explain to my second-grade class what had happened. I had no time to give her actual lessons to teach, so I just wished her luck! I next phoned as many relatives as I could on my cell phone, as none of them could get through on our house phone. "We take CNN first," Jack said. "Then *The New York Times.* I'll be setting up the phone interviews as they come in." Each time Jack picked up the phone, he would say: "Laughlin residence." I heard these words spoken at least a hundred times in the next several hours. I informed him that I wanted to give Bob a break on the hour, each hour, and Jack faithfully granted the request all day long. It was impossible to do a reality check otherwise. Jack and Bob got along well, with Jack exclaiming: "He's got a great laugh! This is a lot of fun. He's good on his feet, really good!"

By 8:00am, every major news station had run the story and everyone finally began to leave our house. By 9:30, things had calmed down considerably. Earlier at 5:00am, I had phoned my parents and my sister's family in Santa Rosa and invited them to come down for the day. Just as the reporters left, my parents and sister and her family appeared with carnations and champagne in hand. Bob's mother would also join us from San Jose. I think it is fair to say that none of us was surprised that Bob had won, but we were, nonetheless, stunned.

A 10:00am press conference had been called on the Stanford campus. We went over by car and Bob traveled by golf cart with Jack merrily seated at the wheel. Bob looked a bit haggard but perked up as several faculty members greeted him as he arrived. We were escorted into a small room with a Stanford banner in the background. As Jack had predicted, the room had two potted plants, a blue wall, and a lectern. It was packed with reporters trying to set up tripods for their cameras. Senior scientist and great friend Ted Geballe arrived to give hugs to all. Friends Sandy Fetter and Mac Beasley came—two men who had been recognized for attracting high-caliber physicists into the Stanford department, including recent Nobel laureates Steve Chu and Doug Osheroff. My family and Bob's mother, Peggy, squeezed into the room as Bob was introduced by President Gerhard Casper. Casper stressed that this was an impressive international prize and that Stanford University was thrilled by the news. He also said that the University of California at Berkeley gave its laureates private parking spaces, but that Stanford hadn't been able to find enough spaces for all its winners! Questions from reporters followed.

We returned home, while Bob was golf-carted off for more interviews. We caught up with him again for an impromptu afternoon lawn reception at Stanford given by students and faculty. Bob's sister Margaret and her husband, Tim, came to join us from San Mateo. At this reception, Bob was essentially "roasted" by graduate students and faculty. He was not allowed any rebuttal time for the roasting. In honor of the occasion, Bob's now infamous pot roast question was retold. He had invented a question for the qualifying exam for Ph.D. candidates that went something like this: "What happens to you if you are hit by a pot roast hurtling through space at 35,000 miles per hour?" He had created it to be a shock-wave calculation problem. To his utter surprise, no one taking the exam got the answer. At the end of the exam, a flustered Russian student approached and asked sullenly: "What is this pot roast?"

"A piece of meat," Bob replied.

"Oh. I needed to estimate its mass."

"Yes."

"I brought my integral tables to this test. I should have brought a dictionary."

Also at this reception was Steve Chu, who passed a eucalyptus branch to Bob as a symbol of "passing the torch."

By 6:00pm, Bob's brother John, wife Maria, and their two sons appeared bearing four pans of lasagna, rolls, and salad. We feasted together and ate a chocolate cake with "Congratulations Nobel Prize!" scrolled in icing on the top a dessert given to Bob from his graduate students. During the day we received glorious floral arrangements from Bob's cousin Libby Hubbell; Visalia friends Betty and Jim Sorensen; my family—the Perrys; former student Martin Griter; and from Texans Bob and Anne Bass who endow

Bob's academic chair at Stanford. I had also spoken to my brother Ned on the phone and he reported being quite moved by the announcement. All of our relatives had already expressed a desire to go to Stockholm with us!

CHAPTER 2
The October-November Diary

CTOBER 14

Still on overdrive, Bob rose at 4:00am and began to check his e-mails. There were several hundred. He meticulously copied each address and answered them all together. Our friends the Esbers wanted to give a party in Bob's honor and Peggy Esber came by to get a guest list with phone numbers from me, which I had gotten together between 5:00 and 6:00am that morning. Both Bob and I returned to work and were beginning to realize that focusing on our jobs was going to be a challenge. We had averaged about five hours of sleep between us and it was a struggle to get through the day. We realized that life wasn't stopping to allow us to assimilate the news of the day before or to plan for our Stockholm visit.

At 6:00pm Bob ran into the house sweating and ranting: "I've really done it now! Edward Teller invited me over to his house at 5:30. I got out of the faculty meeting, got on my bike, got to his street on campus, and forgot the house number. None of his neighbors knew or would tell me where he lives! He's unlisted. He's going to rip me to shreds!" This was the same Edward Teller who worked at Lawrence Livermore Labs in the early 1950s and was the father of the H-bomb. I sat Bob down, we tried to think of people who knew Teller, and finally called a Livermore friend and got his phone number. Bob called, apologized, and asked Teller for his house number.

"I don't know the house number," Teller replied. "Wait and I'll get it."

"He doesn't know his house number?" I gasped.

"That's not surprising," Bob said.

"It isn't?" I countered.

When Teller finally found his house number, Bob offered to bring him a piece of Nobel cake. Teller politely declined. Bob left again and was relieved not to have angered Teller. He and Bob had both worked at Lawrence Livermore Labs in the early 1980s and on one occasion Bob was one of a group of physicists who were summoned to give audience to an experiment Teller designed having to do with using beryllium metal to look for Bloch oscillations. Bob explained to me: "These were oscillations that occurred when electricity was introduced into a solid, causing the currents to wiggle rapidly. Theoretically, this wouldn't happen unless the electricity was applied violently with huge electric fields."

Teller generally summoned people, rather than ask for direct critiques of his work, and this was an uncomfortable forum for Bob who valued scientific integrity above all else. At the meeting, everyone expressed encouragement for the experiment, except for Bob who rigorously and bluntly objected, saying: "It simply will not work." Bob's objection was based on the chemistry of metals that it was impossible to make the beryllium material pure enough. This was a highly complex experiment and it was just not clear that it could be done. To the best of Bob's knowledge, now thirty years later, it has still not been done. Long afterward, Bob's boss told him that Teller had become so angry that he'd asked to have Bob fired. Bob's boss ignored the order, and Bob eventually left on his own in 1985 to join the theoretical physics department at Stanford University.

The meeting at Teller's house turned out to be a cordial one, as Teller seemed to be offering his apology. The Livermore incident just described is demonstrative of Bob's tenacious confidence in his own scientific intuition and knowledge. In the early 1980s, Teller clearly did not appreciate the unbridled opinion of a junior scientist, even though Bob had just quietly published the paper that would ultimately win him a Nobel Prize.

OCTOBER 16

The reality of the prize had begun to sink in. Bob's colleagues responded with e-mails and handwritten letters from all over the world. Physicists who were conferencing in Hawaii, including Klaus von Klitzing, all graciously signed a congratulatory plaque. A unique five-foot scroll of Chinese calligraphy was painted in Bob's honor. His office filled to the brim with colleagues and students all eager to congratulate him with handshakes and vintage bottles of scotch. This was especially welcome since an aura of fierce competition existed at the time as professors struggled to get National Science Foundation grants and other types of funding to support their graduate students and their research projects. Still, a Nobel Prize was certainly the ultimate recognition and generated positive vibes for its recipients for the most part from colleagues. While it was not obvious that the prize would later help in the grant approval process, there was no question that Nobel Prizes in the Stanford physics department were multiplying like fruit flies.

At my own public school, located on the Stanford campus, my students were too young to understand the significance of what had happened.

One student asked "Was it like he carried that torch thing in the Olympics?"

"Not exactly," I replied.

"What was it he got? A trophy? I got lots of soccer trophies small ones, giant ones! I got baseball trophies too."

"No, he got a gold medal for explaining a science experiment."

"I can do that! What did he explain?"

"I'll have him come in and tell you about it," I promised. Bob did eventually make a visit to the class, but he somehow got into the subject of black holes and robotic vacuum cleaners instead. My students were startled as the classroom filled with other teachers and parents, all eager to hear the particulars of the prize.

How did we hear about it? When was I going to Stockholm? What was I wearing? Could I take guests? Who would teach my class for two weeks in December? Would I be leaving lesson plans or would the substitute be on his or her own? Who was the substitute? That's a lot of lesson plans! Was it the Peace Prize he won or the Physics Prize? What about our teacher-parent conferences? Would they be canceled or what? Is there going to be homework while you're gone? My child isn't crazy about substitutes...

"Was that *your* Bob I heard about on the radio?" asked Principal Gary Prehn.

"Yes, and I think I need two weeks off in December," I blurted out.

"I'm sure no one in the district would object. Congratulations!" he replied.

After a frenzied day at work, we all piled into the car that evening to drive to Peggy and Ed Esber's home with family members following behind in the dark. In our car,

Todd and Nathaniel still seemed stunned and uncertain about how to react. One of them commented: "Guess I don't have to prove how smart I am anymore," to which I replied, "Yes, you do." Both boys seemed distracted and disinterested in their schoolwork, and they were to experience lower grades that semester as a result. Their difficulties were not helped by their teachers, several of whom refused to give them make-up work during the week they spent in Stockholm. To this day I find their unhelpful attitudes baffling.

When we arrived at the Esber's, there was a guest-book in the grand marble entryway of their home along with photos of Bob's high school swimming and water polo teams. The house soon filled with nearly one hundred Stanford colleagues, family friends, and relatives. Nobelist Doug Osheroff and his wife Phyllis came to the party and gave us a few tips about what to expect. I recall Phyllis saying in a hushed voice: "You'll need a couple of long dresses. Don't worry. If I could do it, you'll be fine!"

We watched a portion of the video documentary that was made about Bob when he won the Franklin Prize in Philadelphia on April 30, 1998. The certificate for this prize reads that he received it for his "creative theoretical formulation of the fractional quantized hall effect which identified a completely new quantum state of highly corre-lated two-dimensional electron gas in a magnetic field." Bob was asked to say a few words and he spoke about the one thing he hoped people would go away with: "a belief in the magic of the world and the ways in which physics plays a part." This comment reminded me of a wonderful handwritten note we had received from Dr. Randy Weingarten, a Palo Alto psychiatrist and the father of Kai, a player on Nat's Little League team. Randy wrote: "Who would have thought spending all those Saturdays watching our sons in Little League, sharing the joy of it all that you were silently and good-naturedly changing the way we conceive of matter and the universe."

OCTOBER 17

Phyllis Osheroff's comment had given this L.L. Bean girl a wake-up call: I realized I had no idea what the dress code expectations were for Stockholm. I needed to find out, so I invited Sheila Wolfson over, a neighbor who had attended the Nobel events as a guest the previous year. She brought over a video of the ceremony in addition to an album she had made of the festivities. I had my mother-in-law watch with me and after several "Oh, my Gods!" we hustled over to the Nordstrom's parking lot. Californians are not known for their formality in fashion. It is a rare day you see a tie at a restaurant, and many Californians are in shorts and flip-flops twelve months of the year. I was determined to tackle the dress situation first, and tackle it I did. Oddly enough, I had recently received a clothing catalogue with a cover featuring a red skirt with black velvet embroidery, paired with a black velvet top and a red satin shawl. I had shown the cover to my sister-in-law, Margaret Martin, one week before the prize announcement, jokingly telling her: "Bob likes red, so this is the dress I would wear to Stockholm." She was as amazed as I was when Stockholm actually called. This was, in fact, the outfit I wore to the Nobel ceremony.

Sheila advised that I take three formal evening gowns. She said the King and Queen of Sweden wanted people to wear bright colors to the Nobel events. Presumably this had to do with the short days and long, gloomy nights one experiences in the winter in Scandinavia. So, one down and two gowns to go, I visited a Thai silk fabric store and decided to have two bolero-style jackets made that would be worn over two sleeveless silk or taffeta gowns. This was a conservative route, but also a creative one as the fabrics were gorgeous—one turquoise with gold, the other white with golden yellow details. What followed next was a series of frenzied fittings with a dressmaker to design and fit not only the gowns, but also the proper petticoat full-length petticoat to wear under the gowns. A formal gown needs to swish when you dance in it, hence the need for poofiness and sensible shoes. The insatiable need to scratch your legs beneath these slips is beside the point. The gown must swish. I am reminded here of another slip story that involved the wife of a laureate who arrived in Washington, D.C. on a Sunday evening for a White House reception and a dinner at the Swedish Embassy. She was going to wear a black lacey cocktail dress but discovered that she had left her black slip at home. The dress was unwearable sans slip. So, high-level back-alley calls were placed, a D.C. department store was opened up, and a slip was brought to her hotel room. These last-minute undergarment arrangements are an indication that "this is your fifteen minutes of fame" but next time, pack the slip. Undergarment trauma could be the subject of another entire book. Suffice it to say that most, if not all wives of Nobel laureates and perhaps female laureates themselves all experience a sort of post-traumatic stress disorder as pertains to undergarment procurement and wearing during an entire week of formal Nobel events. This is, however, a small price to pay for the privilege of attending the events themselves. A thoughtful and informative pantyhose discussion will follow in a subsequent chapter.

OCTOBER 18

Again landing at Nordstrom's, my mother-in-law and I had lunch to calm our nerves, followed by the impulsive purchase of two Phantom of the Opera black velvet capes with hoods. In retrospect two fur-lined black velvet capes with matching L.L. Bean fur boots, mittens, and mufflers would have been a more sensible investment. Little did we know we'd be met with a blizzard upon arriving in Sweden.

"Good. We needed these capes," we said simultaneously.

"What is it we need again?" Peggy whispered.

"I don't know," I replied.

There was a pause. "You can't go wrong with black skirts, pants, and a silk blouse," she said wisely.

"Right," I agreed and off we went to Macy's to look around. Did we need to confer with a personal shopper for advice? Yes. Did we know what to ask for? No. We decided to go for conservative chic, whatever that was.

OCTOBER 19

The time had come to itemize the things that needed to be done in advance of our Stockholm trip. I needed to feel some semblance of things being under control. Clearly they weren't, but lists always calmed me down. Bob and I were both moving at a manic pace from dawn until late at night. Our main concern was getting enough tickets for the Nobel banquet and the ball that followed. Was there a guest quota? The Nobel organizers had suggested that ten guests was the expected number. Bob and I had a total of thirty-four to consider: our siblings, their spouses, and our nieces and nephews. How many family members were actually going to attend? I phoned my siblings and Bob's family to explain that certainly they were welcome to come, but that I didn't want them flying halfway around the world until we found out what events we could get them into. Bob reserved eleven seats on SAS when an agent called from New York to inquire how many he would need. Six days after winning the prize, this was our best guess. Things were moving at lightning speed and the daily decisions to be made were overwhelming. Bob was asked to submit photos of himself to Stockholm. He ran to Kinko's to get some. There was also a request for an autobiography.

"You're only forty-seven," I said quizzically.

"I'll think of something," was his speedy reply. He began to write and eventually produced a stirring autobiography, now posted on his website.

October 19 was the day Bob officially accepted the award, which I calmly pointed out he had forgotten to do via e-mail. Minor details tumbled over major ones, and we still needed clarification about many things. At the time of our wedding, my parents lived on the East Coast and Bob and I were to be married in the Stanford Chapel on the West Coast. I wondered aloud how my mother had managed to mastermind the intricacies of that event from three thousand miles away.

"She had that Invasion of Normandy chart," Bob replied dryly. "You know, that four-foot by four-foot graph of every conceivable detail and potential glitch. I bet she still has it."

Graphing in minute detail is a genetic trait I share with my mother, so I did color-coded graphing on a wall-sized chart: *Events. Who? What? Where? Contact numbers. Clothing requirements (men's white tie and tails, shoe and shirt sizes). Travel plans. Transportation from the airport.*

Hotel plans. Hotel contact numbers. Problems I know about. Problems I don't know about. Special guests to invite. Food allergies. Palace protocol. Palace protocol?

At some point, it was imperative to step back and breathe. This was beyond overwhelming. The second best way I know to calm down is to buy a pair of shoes. This I did, ordering gold flats to wear with my gowns. I envisioned long periods standing and decided flats a better alternative than pumps. What I hadn't thought about was how slippery new shoes can be, especially in combination with polished parquet palace floors. More on this fashion faux pas in the "Dining with a King" chapter.

While I was busy graphing and trying to go to work, friend Peggy Esber took my mother-in-law shopping for evening gowns. Bear in mind that there are evening gowns, and there are evening gowns. Trying to determine whether a gown is chic, comfortable, sufficiently formal *and* the right color and style is an art in itself. Peg Laughlin ended up with an elegant purple gown with a jacket—purple, the color of wisdom.

Later in the day, Jack from the Stanford News Service delivered some audio and videotapes of Bob from October 13, a day that he apparently talked for twenty hours straight!

OCTOBER 22

This had been an exhausting week for Bob and me. We'd had little sleep and much anxiety over how many tickets we'd have for various events. We did receive a very helpful e-mail from Bill Phillips who had won the prize in 1997. He gave us all kinds of practical advice, especially about having patience with the ticket situation. I recall his opining that he wondered where all the "physics groupies" were when he won his prize and his wife Jane retorting, "Bill, you wouldn't know a physics groupie if she was standing in front of you!"

Some years later, Jane and I met socially at a Lindau Nobel Prizewinner meeting held at Lake Constance, Germany, and became mildly hysterical over outerclothing and underclothing issues, shoe travails, and protocol challenges. We wondered half-seriously why some sort of Swedish Survival Kit for Gown Beginners (SSKGB) is not issued to laureates and their families via FedEx the day of the Nobel announcements. That conversation was actually the genesis for this book.

This particular morning we finally received notification about being assigned additional seats for the Nobel ceremony! Things looked good. Now all that was needed was a finalized list of our guests. In the meantime, Bob packed his things for a flight to a conference in Florida, that was leaving at 5:00am in the morning.

OCTOBER 23

Bob arrived in Tallahassee for his weekend conference. Fellow Nobel co-winner Horst Stormer was there and Bob was thrilled to see him. Horst and Bob were often

39

seen in various states of convulsive laughter when together, and this time was no exception. Horst told Bob that at an earlier reception, a young man had come up to him and said: "Oh, so you're Horst. My thesis advisor told me all about your work. It's fantastic. I bet you win the Nobel Prize for it someday."

Horst said he smiled slightly and replied: "That good? Yes, maybe you're right."

While Bob was in Florida, I continued working at home graphing and issuing invitations to Stanford professors, former colleagues, and family, hoping to receive some confirmations. The final list was due to Stockholm by November 6 at the latest.

OCTOBER 24

The day began with a violent thunderstorm. I spent the morning at the Thai silk store throwing bolts of fabric across tables, waiting for further Divine intercession. Receiving none, I returned home and spoke on the phone to Sandy Fetter. Bob had called Sandy earlier in the week because Sandy had been hospitalized to have a pacemaker installed beneath the collarbone. Bob had asked him (moments after surgery) to come to Sweden with us. I warned Bob this wasn't the kind of news a guy with a three-hour-old pacemaker needed to hear, but he had called anyway. Sandy was a bit groggy but told Bob he'd be delighted to come along with his friend Lynn. Today's phone call from Sandy was to reconfirm that he had not been hallucinating and that Bob had actually invited him. I reassured him and said I was pleased to know he'd be joining us.

I spent the afternoon with my dressmaker who is an energetic woman with a houseful of children. Her eldest met me in the driveway, covered head to toe in mud. He'd been playing football and had invented a way to power-spray every living soul in a five-block radius with a flick of a switch. Dripping wet, I greeted Barbara and we discussed the design of two gowns and jackets, their petticoats and the golden slippers. There was still a surreal element to all of this, as though another Anita was having these experiences. My life had become prioritized around these seemingly mundane chores, which were collectively only my best guess for what was appropriate for a series of occasions over which I had no control.

OCTOBER 30

Mats Johnson arrived at our house in the evening. He was the man who had presented Bob's case before the Swedish Academy members the people who ultimately meet to decide who will be the physics winner. This involves a long, secretive process in which the credentials of those nominated are scrutinized by Swedish physicists and others familiar with the theories or experiments being examined. Mats told us that the Physics Prize was thought to be the most prestigious of all the Nobels. He also told us to expect cold weather, grand evenings, and limo service with a driver who knew how to navigate Stockholm and who would be available to us around the clock. We would be assigned to an attaché, a person to whom we would stick like glue from the moment of our arrival until our departure. The attaché would give us "The Schedule." Mats had a wonderful rapport with Bob and his gentle manner somehow calmed us both down. We took him out for champagne to celebrate.

NOVEMBER 1

Bob's forty-eighth birthday! We celebrated with pumpkin pie and Bob's mother gave him a book on teenagers and dating that our younger son, Todd, grabbed and took up to his room. It hasn't been seen since. Over breakfast Bob sat me down and said there was something new that had come up that wasn't on my master chart.

"I don't have room for anything else," I snapped.

"It's an invitation to dinner."

"That's nice." One plain old dinner I could handle.

"To the Swedish Embassy in Washington, D.C."

"The what? In Washington?"

"The Swedish Embassy and there's more. We go to the White House first for a private reception with the Clintons. Not formal, but probably formal. I think."

"I see," I responded, "and when is this?"

"On November 24, Thanksgiving week. All the American winners and spouses are invited. We *even* get to pay our for our own plane tickets to get there."

Apparel trauma set in once again. "Formal or not formal" at the afternoon White House reception and probably "more formal" at the embassy dinner meant what? I guessed that I needed two separate outfits. It meant a new suit for Bob and a tuxedo rental for the evening. Bob's mother had a two-word reply: "Beaded jacket." So I set off to find a black-and-white dress over which I could wear a black beaded jacket for the evening. I was on apparel overdrive. Bob, meanwhile, biked over to the tuxedo rental store to rent a tuxedo for Washington and to get measured for white tie and tails for our Stockholm events. He took Todd and Nathaniel with him to get their measurements. We were to eventually accumulate a two-inch binder full of clothing inquires in addition to the measurements of each male family member and guest, which Bob dutifully e-mailed to Stockholm so that the correct white tie and tails would await them on arrival. It looked like our family and guest list was approaching thirty people.

NOVEMBER 4

This morning a man who was producing a Swedish documentary arrived along with several freelance photographers and set up shop in the middle of our living room.

They interviewed Bob, and then recorded him playing one of the five original piano sonatas he had composed. Bob's musical abilities have always been intertwined with and complemented his ideas in physics. He bounces back and forth between piano bench and desk chair when working at home. The producers told him the viewing audience for this documentary would be around 700 million. Luckily Bob had finished playing his last sonata before hearing that comment. They followed Bob to his office, where they photographed him with students. Later they got a photo of Bob talking to Sandy Fetter on campus in a eucalyptus grove. The producers liked Bob's idea of using glass itself as a visual image and shot a picture from a fourth-floor window on the campus. Later, still-water shots were also incorporated. Bob returned home early and though exhausted, sat at his piano playing for several hours.

NOVEMBER 7

With our guest list finally submitted, we had a temporary feeling of things being under control. We invited Bob and Anne Bass over for champagne. At the time, Bob Bass was the president of the Stanford trustees. In addition to sending us a lovely ivory floral bouquet when the prize was announced, Anne and Bob later gave us lovely Steuben crystal apple, suggestive of the Newtonian apple. On this particular evening, they graciously treated us to dinner at Spago's restaurant in Palo

Alto. Bob Bass later recalled: "You [Bob] called me to report your research on what to take to Stockholm and included that Anne should take 'in' dresses or 'inn' dresses. It took me some questioning to learn that you were referring to 'n' dresses, as the variable."

NOVEMBER 14

Today we met our attaché, Aja Lind, via e-mail. The attachés for the laureates are an elite, hand-picked group of individuals who are multilingual and multi-versed in the nuances of Nobel protocol. A lifeboat at last!

As a first request, we asked Aja to make dinner reservations for the various groups of people accompanying us, different groups for different nights, specifically for those occasions that we could not join them. This of course required a trip back to my master chart to see who was where when and needed dinner plans, as we wanted everyone in our party to stay connected. Our guests now entailed approximately twenty-six people coming to Stockholm from all over the United States. We asked Aja to choose the restaurants herself. I also asked her to make me some hair appointments with her favorite hair salon in Stockholm. We now had clarification about numbers of people we could invite to what events, so an additional chart was created to place guests at various events to give everyone a chance to attend some functions with us. This presented a Herculean juggling act, which kept Bob and me up at all hours.

My two evening gowns were finally finished, and next came the task of finding jewelry I could afford to go with them. Apparently I could not borrow a Harry Winston 80-carat diamond necklace with matching toe rings like the Oscar-nominated actresses get to wear. But why not? Certainly the Nobel Prize banquet (to which one had to be invited and that people all over Europe yearned to attend) was as illustrious an event as the Academy Awards. Why would a physicist's wife be less trustworthy than a Hollywood actress when it came to borrowing jewels?

In addition, more suitcases needed to be purchased for the four of us traveling together, and the men in our family all needed suits, shirts, ties, socks, and dress shoes. These purchases were sandwiched in between my half-hour parent-teacher conferences, which were to be completed before Thanksgiving and prior to our flying to Washington, D.C. Bob managed to not only to teach his undergraduate course and grade physics problem sets, but also to nearly finish his autobiography. We were off to Washington the next week with one question on our minds: Given the Lewinsky scandal, would Clinton still be in office?

CHAPTER 3
American Laureates Reception: Washington, D.C.

"PASSPORTS, PASSPORTS, WE MUST BRING PASS-ports!" Bob reminded me. "Can't get through White House security without them. They e-mailed me nine times to bring them!" We flew to Washington on November 23 and taxied to the Watergate Hotel, which was being run by Swiss Hotels. Our suite was breathtaking, with pale-green-and-white striped wallpaper flanking floral curtains. The view was of sparkling water below. When we entered the suite, operatic music was playing, the lights were dimmed, and the bed turned down. We found a plate overflowing with fresh fruit including boysenberries along with mineral water and a fresh bouquet of blooms. We also received a generous gift from the Swedish ambassador, which was an elegant pair of crystal champagne flutes in the same style that we later found on the Nobel banquet tables. Bob uncharacteristically picked up the phone and impulsively ordered a $75 bottle of champagne and some dinner to go with it. The lights twinkled on the water below as we sipped our bubbly.

The next morning we slept late and went down around noon to a Swedish press luncheon that the Swedish ambassador, Rolf Ekeus, had kindly arranged for the laureates. Three physicists, three winners in medicine, and one chemist arrived with their spouses to answer questions from the press. Afterward we sat down to salmon and sorbet. I sat next to Bob's fellow winner, Dan Tsui. Despite a persistent heart condition, Dan radiated happiness and was full of boyish energy. Bob was seated with Horst Stormer and naturally the two of them were out of control with laughter throughout the luncheon, in spite of the efforts of Horst's wife, Dominique, who had positioned herself between them as a buffer. It was at the end of the luncheon that we were handed an "updated itinerary" from the White House. The Clintons had requested "we not dress formally." The wives looked at one another with some apprehension, and then couples returned to their rooms to spend the rest of the afternoon arguing about what to wear.

At 5:45pm, we all congregated in the Watergate lobby and got aboard a mini-bus. Ambassador and Mrs. Ekeus limo escorted us directly to the East Wing of the

White House. When we pulled up to the guardhouse, the bus engine was turned off and a guard dog circled the bus for ten minutes sniffing for bombs. Apparently cleared, we got off the bus, entered the guardhouse, showed our passports, and walked through a metal detector. Horst Stormer searched his coat pockets. He had forgotten his passport. Dominique had remembered hers. Nine times we had been told to bring our passports. Horst endured a bit of hazing and laughter from the other laureates until the guards relented and let him in. After all, his name was on the guest list.

We all proceeded inside, up some stairs, and down a red carpet to a reception room at the end of the hallway. A large portrait of Lincoln hung in the room and in front of it played a Marine string quartet was playing in front of it. The surrounding floral arrangements were gloriously artistic and fragrant. The round table in the center of the room had containers of iced tea, bottled water, and greasy pastries on it, which I decided was a misguided American attempt to give an English high tea. We were invited to wander through all the rooms on the floor. Stationed at each doorway, we found military personnel who explained the history of the room, furniture, and portraits. I was fascinated by the official portrait of Jackie Kennedy Onassis, in which she appears like an apparition in her ghostly white gown, gazing into the room and beyond. Outside the windows, the Washington Monument seemed to walk with us from the Red Room to the Blue Room to the Green Room, ever vigilant.

Shortly after we arrived, Hillary Clinton entered the reception hall alone. She is petite in stature and her demeanor with the male laureates was cordial. She told Bob she had just returned from a Central American tour and Bob replied that I had a relative in the Peace Corps stationed in Haiti—who said that the country had become a desert, an ecological disaster area. She agreed, saying it could be observed from the air. After speaking briefly with each laureate, Hillary ignored the women in the room and seemed to vanish into some invisible White House portal on cue as Bill Clinton entered the room. The President bounced from guest to guest making everyone comfortable, smiling and thanking the laureates for their contributions as he met them. He made a short speech quoting Bob directly from a Stanford alumni magazine article in which Bob had said he owed a debt of gratitude to his parents' generation for funding science and making his discovery possible. Clinton offered the comment that if the government does not continue to make similar commitments to investment in scientific research, our children's generation will suffer. Afterward when Clinton stepped off the podium, he found Bob and me and put his arm around my shoulders several times to pose for pictures I have yet to see. I was wearing a voluminous, square-shouldered, camel-colored wool blazer with a pleated skirt, which unfortunately made me look a little like Eleanor Roosevelt from the rear. Undeterred by my suffragette profile, Clinton was friendly, youthful, and unpretentious. He was someone who would gladly give you a lawnmower from his garage if you needed it. His informality with people is a true gift. We stayed for an hour and a half talking with other guests such as the head of NASA, the surgeon general, and Clinton's science advisor, Artie Bienenstock, who lives across the street from us now at Stanford.

After being herded past bomb-sniffing dogs and back onto our mini-bus, we were rushed to our hotel and given twenty minutes to "freshen up," which translated into ten minutes of elevator rides, five minutes of ripping panty-hose, and five minutes of swearing. This task, as you might imagine, was Mission Impossible. Had we known about this Houdini act ahead of time, we all would have put on our evening clothes and worn trench coats to the White House.

Back onboard a second time, we hyperventilated toward the stately home of Ambassador Ekeus and his wife. It was located next to Al Gore's home. The driveway was as full as a car lot and we seemed to be the last breathless guests to arrive. The impressive home was packed to capacity and we literally squeezed our way through the front door. The guest list read like a Who's Who of Washington politics (included in the appendix). Given a glass of champagne, we were led by the ambassador directly to Chief Justice William Rehnquist who silently towered over everyone with the wizardly mystique of

Professor Dumbledore from *Harry Potter*. Bob recognized Joel Klein next to him and remarked to Joel: "Oh, so you're Joel Klein. Very happy to meet you because this Microsoft suit you're engaged in is profoundly important."

Klein turned to Rehnquist and said: "You know this guy isn't as dumb as he looks."

Klein was the federal prosecutor in the Microsoft anti-trust case, which he eventually lost. Like most professors, Bob felt that communications standards should not be monopolized. This is one of the issues that later motivated Bob to write his second book *The Crime of Reason* (2008).

Next to Rehnquist was Justice Ruth Bader Ginsburg who thoughtfully asked how we were holding up on this Washington trip, just as a tsunami of guests began to rise and crash in our direction, depositing us into the dining room before I could reply. A laureate was seated at each table, and the guests were asked to write down three questions per table for each of the recipients to answer later in the dinner. Quickly seated, we were introduced to others at our table. They included an animated Michael Heyman, who had been the chancellor of U.C. Berkeley and was then director of the Smithsonian Museum Institution. As a Berkeley alum, Bob discussed the old Berkeley days with him. We sat with members of the Swedish press and with a Swedish mining expert whose job it was to teach Bob and me how to toast in Swedish at parties by saying: "Skoal!" He made sure we rehearsed this toast with gusto at least forty-three times throughout the evening. Seated to my right was a soft-spoken, intense gentleman who turned out to be *New York Times* columnist William Safire. He had been the former press officer in Richard Nixon's administration and is a brilliant conversationalist. When I told him I was a teacher, he replied: "Oh, you're at the front lines then," a comment that made me pause. He had a stream of questions for Bob and discussed emergentism, a subject that later coalesced to form the core of Bob's first book called *A Different Universe* (2005). As Bob explained: "The central idea of emergentism is that much physical law results from organization, the way the meaning of an impressionist painting does, and so ceases to exist when you take it apart to see how it works."

We enjoyed a delicious meal of salmon and capers, beef and potatoes, with ice cream in a pastry shell for dessert, followed by brandies and liqueurs. There was a question and answer period that followed. Bob was asked: "Do you think we'll soon have a unified field theory?" This was an oblique reference to Einstein's thinking about the cosmos.

As I recall Bob's response to the question was: "No, such theories cannot be experimentally falsified at the moment and are thus pure science fiction. Period." A hush descended over the room.

"Are you sure?" I asked as he sat down.

"I'm sure."

CHAPTER 4
Traveling to Stockholm

ONE INCONVENIENT FEATURE OF TELLING YOUR extended family, out of the blue, that you've won a Nobel Prize and they could come celebrate with you in Stockholm, is that while there is initial emotional shock, it is followed by an even larger sticker shock. Shockwaves seem to follow my husband. In an effort to defray the up-front costs of the travel to Stockholm, guests of the laureates are allowed to charge some of their expenses to the prize money account if authorized in advance by the recipient. This is only fair, unless your relatives have actually banked the money ahead of time in feverish anticipation. It is no doubt an accounting headache for the

Nobel officials, but they do everything in their power to help recipients with great decorum and tact.

On December 4, we departed for the San Francisco airport in a stretch limo complete with a neon-lit disco bar, which was not especially appealing at 6:00am. This limo had been Todd's idea, but unfortunately he slept all the way to the airport. We carried with us every suitcase we owned and I was careful to let the boys do all the lifting. The day before we left, I had my nails painted for the first time since I was twelve and had used drops of food coloring which created permanent finger stains that lasted for weeks. Now, Rose, my manicurist at LaBelle Spa, promised that my bright ruby red nails would hold up for two weeks if I didn't wash dishes, zip luggage, or scratch excessively. These things I avoided like the plague.

Arriving in the domestic terminal of Chicago's O'Hare Airport and trying to locate the international terminal was a hair-raising ordeal, even for people as bright as we all thought we were. It was the blind leading the blind, and once the SAS gate was located, my seventy-three-year-old father celebrated his five-mile jog through ten unmarked terminals by bringing a Chicago hot dog to the gate.

"You can't take that on the plane, sir. It needs to go by itself through x-ray," the security guard cautioned.

"Why is that?" my father asked ferociously.

"All hot dogs get x-rayed, sir."

So as a dutiful American citizen, my father laid down the dog and walked next to it as it made its way down the conveyor belt. Moments later, with ketchup and mustard oozing out ahead of it on the belt, Dad retrieved his hot dog, and off we dashed for the plane.

Bob and I were assigned luxurious first-class seats while our relatives were all happily partying back in economy. We thoroughly enjoyed ourselves, were introduced to the captain and Bob signed an autograph for the flight attendant's daughter. No one in the United States had asked for an autograph and it seemed fitting to give a first signature to a Swedish family.

Upon arrival at the Arlanda Airport in Stockholm, we were immediately greeted by Swedish Academy officials, one of whom was Erling Norby, the head of the Swedish Academy. Bob, Nat, Todd, and I were taken to a VIP lounge that had tables full of flowers and food. Glass bottles of Coca-Cola were also available. It was there that we met our lovely attaché, Aja Lind, in person. She is someone with whom we became inseparable throughout our visit and we continue to hold in fond regard. She is a tall, slender young woman with dazzling green eyes and a beautiful smile, and our sons were smitten. Aja's occupation had involved working for the Foreign Ministry office in addition to traveling extensively to set up computer systems in Swedish embassies throughout the world. She was bright, articulate, and communication-savvy, and we were thrilled to have her on our team.

Photographers arrived in short order to take our pictures with a portrait of His Majesty the King in the background. While we relaxed, our luggage was being collected and taken through customs to be loaded outside into our waiting limo. The room livened up several moments later as Horst Stormer and Dominique arrived from their New York flight, along with one of the winners in medicine, Lou Ignarro. All three men posed for group photos, which appeared later that day in the Swedish newspapers. I don't think any of them actually stopped smiling for their entire visit and were great fun to be around. We did all wonder how our luggage was being magically collected for us and I worried that some of mine would be sent back to the States.

Walking out to our limo, a sleek black Volvo with *Nobil-12* on the dashboard, I whispered to Bob that our luggage would never fit. Sure enough, a taxi was parked behind the Volvo with our eight suitcases mushrooming out the side of the trunk. Each Nobel laureate was assigned a car and driver for the duration. Our driver, Lena, seemed calm as the four of us piled helter-skelter into the Volvo. We followed Horst and Dominique to Stockholm, passing the Swedish Academy of Sciences in the dark distance. To our delight, a major snowstorm was arriving with us and the hillsides were blanketed with white. Saying it was a "wonderland" is an understatement, as nothing can describe the beauty of that mid-morning drive, just as the sun was dawning on the glittering snow. We were speechless as our car's sleek black shape maneuvered past white birch forests pencil-lined with snowfall.

Our first view of Stockholm was of dark flowing waters coursing beneath rounded bridges connecting a series of islands. Ferryboats were anchored at docks on the water, unfurling their flags as thousands of white lights twinkled overhead. It was a city pulsating with light, motion, and majesty. The Grand Hotel is located in the center of the city, facing the Royal Swedish Palace and Old Town. It is the oasis for laureates and their families who come each year and though palatial itself, manages to be an inviting and intimate hotel with impeccable service. Bundled in a long black overcoat, I remember being ushered through the front door of the Grand, past lighted evergreen trees. Once inside, we were met by a photographer and the hotel manager who escorted us to our rooms. We stayed on the fifth floor with the two boys on our left next door, Peggy to our right, and other relatives farther down the hallway. All of our rooms looked out onto the palace, the swirling blue-gray water, and thousands of white lights still illuminating the city at mid-morning. Our room had seven-foot-high windows that opened grandly onto the scene below and we opened ours to let tiny snowflakes billow through the room like fairy dust. The suite had fresh flowers and a huge fruit platter with assorted chocolates. We had filled out a food preference and allergies survey for ourselves and our guests before coming to Stockholm. Those preferences were carefully monitored during our entire stay. Aja had made sure the boys had a large basket of hard candies in their room when they arrived and even tried to install a personal computer for them.

I began to unpack formal clothing first, as Bob was escorted downstairs by Aja to view the plans for the week's activities with her. She gave Bob a large computer print-out on which she had a weeklong daily schedule for us, broken down by the half-hour, beginning at 8:00am and ending around midnight. Too tired to be fazed by this document, Bob and I went to sleep and awoke at 7:30pm, discovering that a further dusting of snow had muffled sounds coming from the streets below. The snow resumed and did not stop for four days. We were in heaven, but it snowed so many feet in towns north of Stockholm, that a state of emergency had been declared. Apparently, snow in Stockholm, was quite unusual but it gave the city a fairy-tale quality that made this Nobel experience all the more captivating.

The snow was shimmering on the waters below at 10:30pm when we heard a knock on our door. It was Bob's mother, Peggy, who came in bearing a half- bottle of white wine to celebrate. Her joy was contagious. We ordered dinner and chatted until I could not stand looking at that lovely snow without walking in it, so arm in arm, out into the night Peggy and I went. In a Swedish winter, it is daylight from 8:00am until around 2:00pm, when children return from school. The rest of the time it is pitch black. In overcoats, boots, and gloves, Peggy and I made our way to the right of the hotel and walked toward the Opera House through several inches of snow. We walked close to the palace perimeter where sculpted stone cupids were romping in snow. The multitude of white lights above illuminated the slippery streets and every window as if to flood moonlight into the city. The jet-black street lampposts stood in bold contrast to the gentle white snow that lined every lamplight and twig. Walking across the bridge from the palace, we paused halfway to look into the water below. Peggy exclaimed: "Who would have believed I would be out walking in the snow, in Stockholm, at midnight and my son has just won the most prestigious academic prize in the world? Who would have thought?" Indeed.

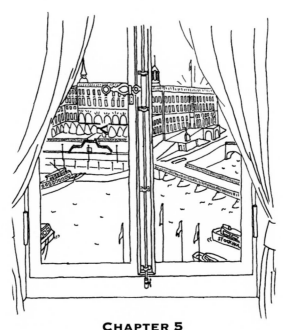

CHAPTER 5
Writing the Toast for the Nobel Banquet

𝕯ECEMBER 5, IT WAS STILL SNOWING IN STOCKHOLM and the quiet had lulled us all into a peaceful slumber. This was to be our free day of exploration in the city. Todd and his grandfather decided to remain in the hotel and use the spa facilities, specifically the fitness center and the sauna. My father later reported to us he was told that the sauna was blissfully co-ed, he'd sat in it for what seemed like hours waiting for a tall blonde to join him. Meanwhile, Nat, Peggy, my mother Carolyn, and I headed off through the snow to Old Town, which meant a treacherous walk across the bridge and past the palace on icy sidewalks. We brought Nat in case one of us should need to be carried back. Old Town was having its annual Christmas Craft Faire. It actually took place on some of the original cobblestone roads that converged into a plaza on which simple wooden booths were erected. Inside each cozy booth, vendors sold a variety of goods such as wooden spoons, straw tree ornaments, brooms, canes, St. Lucia dolls, knitted hats, socks, and mittens. Helping ourselves to some cups of hot and potent Swedish glogg (which resembles American hot vodka punch), we felt fortified and shopped. I purchased a set of wooden candlestick holders for Peggy, Carolyn, and myself that were beautifully painted in soft colors. We all three still get these out every Christmas to remember our Nobel week. Sipping more punch, this time with peanuts and raisins floating in it, we headed off to have lunch, toured some art galleries, and best of all, discovered a Scandinavian sweater shop. It was snowing, and we were in Sweden, and though I would need to take it home wrapped around my head like a turban for lack of suitcase space, I purchased an exquisite cardigan of blue and white wool. I have worn it many times since then and like a best friend, I always bring it back with me to Stockholm when we visit.

Back at the hotel, we woke Bob, Grandpa, and Todd from their four-hour nap and set out once again for Old Town to have dinner in a restaurant Nat and I had seen earlier in the day called the Dragon Palace. This was an intimate Chinese restaurant with mahogany and ivory carvings on the walls. There was so much greenery, it created a striking underwater effect. We were seated in the corner at a large round table. Bob was feeling discouraged because he had been unsuccessful trying to think of a toast to deliver at the Nobel banquet. Typically one recipient from each of the disciplines is elected to give a toast. Horst and Dan chose Bob. Luckily for Bob, my father was in rare form that evening. A Yale and Harvard man, he is seven-tenths academician and three-tenths Broadway entertainer. On this particular evening, he kept the seven of us laughing for three solid hours. He had Clinton jokes, Viagra jokes (the Medicine laureates had found the cell signaling that had led to the association with Viagra's promising properties), sauna stories, king "opening lines," bald jokes, microscopic mirror jokes, and on and on. Bob couldn't get a word in edgewise. Interestingly enough, we were all drinking beer except my father who was sipping tea leaves of questionable origin.

Asked what he thought Bob should say for a toast at the banquet, Dad suggested: "You've all been sitting patiently for three hours. The men go that way, the women, that way! It's every man for himself!" He was referring to the rumor we had heard that at the banquet, no one is permitted to leave the room until the king departs.

Asked what he would say to the king, he replied: "I'd look him right in the eye and say, 'Just what is it a king does all day?'"

Asked what he thought of his first Swedish sauna experience, he said: "Well, see everybody's sitting close to me now, do you see that? It's the hair tonic I found at the spa—drives women mad."

It was the spirit of that dinner that relaxed Bob enough to finally sit down and compose the toast that he was to give before 1,300 people and millions more via television on December 10. When we returned to the Grand Hotel, I suggested he begin with something from Mark Twain, his favorite writer. My comment combined with my father's string of hilarious observations, got Bob writing and in a short time, he had the toast composed. Like any good writer, he wrote about what he knew, in plain and simple terms. He began by calling attention to Mark Twain's speech about babies. He then cleverly discussed the trials of being a father himself and what having young children can entail. He ended by stating that his toast wasn't to babies exactly, but to all parents who nurture their children. He mentioned that his own father had died several months before the first of his now famous papers was published and asked everyone to raise their glass to parents—both present and gone. When Bob gave his toast from the podium at the Nobel banquet, there was not a dry eye in the room even men seated around me were in tears. Ironically, Horst Stormer's father had died shortly after Horst found out he had won the Nobel Prize. We also knew that Dan Tsui's parents had died of starvation in China. So each of these three men had to deal with the deaths of a father who had nurtured and mentored him into manhood. Their collective presence was missed deeply in Stockholm.

CHAPTER 6
*Dinner at the Swedish Academy of Sciences
and the Nobel Lectures*

𝕵T WAS STILL SNOWING GLORIOUSLY ON DECEMBER 6.
Bob and I took some time to absorb Aja's master-plan
printout, talked about the wording of the toast, and then
had brunch with Horst and Dominique. This turned out
to be our only meal in Stockholm with just the four of us.
We were all having some difficulty sleeping but loved the
seven-foot Swedish bathtubs and down comforters. By
now all the laureates had arrived. The lobby of the Grand
Hotel began to fill with guests from all over the world who
met to chat and enjoy afternoon tea.

Between 5:00 and 7:00pm that day, there was scheduled an informal reception which the laureates and their families were required to attend in a room at the Grand Hotel. The clothing lingo for "informal" in Sweden meant coat and tie for the men, dress or suit for the women. This gathering was a Swedish pep rally of sorts at which we were all welcomed and then shown a short video of the Nobel ceremony, which had several of us crying uncontrollably into our thin paper napkins. Nat and Todd stood tall, quietly sipping sodas, but were both clearly feeling like deer caught in headlights. We were encouraged to ask any protocol questions we had. The rumor about zero bathroom breaks at the banquet was confirmed. No one was to leave the banquet hall for any reason while His Majesty the King was present. The banquet would run precisely three hours and twenty-five minutes, with a half dozen wines including champagne and after-dinner drinks. Bathrooms were located on the same floor and could be used after the banquet and before the formal ball that followed upstairs at the top of the Grand Staircase. I saw my father sit down, wipe his forehead with his handkerchief, and then wink at our boys as if to say: "I'll be all right, don't worry about a thing!"

By December 7, Nat and Todd had quickly acclimated to infinite room service, as evident by the trays of partly finished cheeseburgers that accumulated in their room. There was an excellent daily Swedish smorgasbord breakfast of yogurt, herring, granola, lox, pastries, and boiled eggs downstairs, but they preferred sleeping late and awakening to burgers. Aja and my parents checked in with the boys periodically as our schedule began to fill. My parents and siblings who had arrived in Stockholm, visited the famous Vasa Museum and spent more quality time in the sauna together. On the bus trip returning from the Vasa, my mother's passport was stolen from her purse, an incident she reported to the American Embassy but kept to herself until after we returned to the States. Two young and dashing Swedish policeman came to her hotel room to get more information and one of them actually mailed a Swedish police badge to my brother-in-law who at that time worked as a detective. Travel around Stockholm is generally efficient and feels very safe, so this incident was unfortunate.

The afternoon of December 7, Bob and I were driven to the Aula Magna Auditorium where the Nobel Lectures would be held the following day. The laureates checked the stage setup and asked for the equipment they would need to give their talks. The ceiling of this auditorium is pieced together with thousands of varied lengths of wood. It is extraordinarily precise and elegant in its simplicity. Like clockwork, we were then all transported to the first official press conference for the recipients of the Physics, Economics, and Chemistry Prizes. This conference was more or less dominated by the economist Amartya Sen, perhaps because his field of expertise is slightly more accessible to a general audience as compared with medicine or physics.

From this venue, we traveled to a reception at the Swedish Academy of Sciences. We stood in awe in this cavernous room in which the names of Nobel recipients are officially announced each year. The room is filled with life-sized portraits of famous Swedish scientists and former heads of the Academy. Outside its vast windows, snow coated every twig of every tree with such crystalline clarity that time itself seemed suspended.

From there, we were invited into a nearby room that was painted a fresh Meyer lemon yellow and had grand portraits of Alfred Nobel alongside the renowned Swedish naturalist-botanist, Linnaeus. I was seated between Stig Hagstrom and Erik Karlssen. Mr. Karlssen had headed the Nobel Physics Committee for the last twelve years, and this was going to be his final year. Mr. Hagstrom had been a science advisor to the king for a long time and was returning to Stanford to teach in January 1999. In the course of the dinner, I found out about the selection process for the Nobel nominees which goes something like this:

- In February: 1,000 universities worldwide are sent requests for nominations.

- In May: 350 or so universities have responded.

- In June: the top 30 candidates are chosen.

- In July: the top 10 in each field are identified.
  At this point, five Swedish Academy members review published papers of the nominees in each field.

- ❖ In August: the decisions are made but kept secret.
- ❖ In September: the officials of the Academy vote on the recipients.
- ❖ In October: the winners are informed by telephone by Academy representatives.

Previous Nobel laureates themselves may nominate recipients each year if they wish. It's a laborious process, which leads ultimately to the recipient's initiation into the most exclusive club in the world.

On the evening of the December 7, our family and guests (a party of twenty-one) dined together at a restaurant called Tivoli that Aja had chosen for them. Our own dinner at the Swedish Academy of Sciences included mussel soup, veal wrapped in pastry with vegetables, and mango sorbet. At the conclusion of dinner, dark chocolates, cognac, and steaming coffee were served in the next room while we were entertained by the celebrated flutist Gunilla von Bahr. She played an exquisite piece composed to mimic birdsongs and later gave us one of her CD recordings. We were stunned by the way her music seemed to breathe life into the many botanical renderings that hung on the butter-colored walls.

The snow had stopped falling by December 8. It was on this day that we officially lost track of our sons and relatives as Aja's master plan kicked into high gear. *Her* plan made *my* plan look like child's play. Our schedule began with a luncheon at the Swedish Academy at which we were treated to Swedish pudding, a delicious dish of scalloped potatoes and salmon, smothered in clarified butter.

After lunch we were driven back to the Aula Magna Auditorium at Stockholm University for the Nobel Lectures. The auditorium was packed to the brim with spectators. Dan Tsui spoke first. This was his first public lecture in a long time due to his fragile health, and he spoke slowly in wispy phrases. We all worried constantly about his heart condition, but we took our cues from his wife Linda's face about how concerned we should be. As an aside I will say that now, ten years later, Dan is physically stronger with increased stamina.

Horst Stormer spoke second. Bob was in the audience doodling last-minute changes onto his transparencies. Horst got his attention with a photograph of the three famous Italian tenors—Placido Domingo, Jose Carreras, and Luciano Pavarotti. Horst had transposed smiling pictures of himself, Dan Tsui, and Bob onto those of the tuxedoed singers, dubbing Bob, Pavarotti. Laughter continued (Bob loudest of all) as Horst interspersed his physics lecture with Gary Larson cartoons. One such cartoon was a line drawing of three guys sitting under some trees. A huge apple was about to fall on the head of one of the guys (Newton, we surmised). Horst described this as "the apple from hell about to land on Bob Laughlin's head." Bob could not possibly top that presentation, but he gave a very personable address to the young audience.

All three men were clearly relaxed full of adrenaline and ended the visit by talking with and signing posters for audience members. Because this day was orchestrated down to the minute, everyone left punctually to appear at a prescheduled television interview which went well and was widely broadcast.

The next portion of The Plan was reminiscent of the Washington D.C. Ambassador's party because we were whisked back to the Grand Hotel and given just thirty minutes to change into evening clothes. Things were accelerating. Luckily, I had penciled in wardrobe decisions for myself into every available space on the master plan so that I wouldn't forget what to wear for what half hour. I had even lined the clothes up chronologically in the closet with the heels of shoes pointing toward me for quick entry. The hotel valets would pick up ironing or mending within fifty seconds of our call to the front desk, so speedy garment maintenance was no problem and nothing short of miraculous.

After changing, laureates and their spouses were taken to the University of Stockholm for an intimate dinner with the president of the university, faculty members, and Nobel officials. Notably, there were very few other women present, which made conversations challenging (the x-y chromosome factor again). I told the charming man seated to my left that I personally, was not crazy about MIT as a university for our own two sons, because I felt the training was too narrow, that equal exposure to liberal arts is important too. I explained that our oldest, Nat, loved chemistry and computers, but was also a gifted improvisational jazz pianist. His brother, Todd, played the viola, hit home runs out of the baseball park, and loved physics. It turned out that this man to my left was none other than the former director of the Swedish Technical Institute. He later stood up to give a toast and said: "I have just had a conversation with Mrs. Laughlin in which she said she doesn't think much of MIT and technical schools in general at the moment [much laughter from the audience] because she feels the training is too narrow. This, too, is what I feel and we hope to try to find a compromise for our students." Heaving a sigh of relief, I lifted my wine glass to skoal. As we ate our dinner, we were serenaded by the President's Choir a group of students who sang beautiful traditional Norwegian and Swedish music. Meanwhile, our family and friends were again all dining together, this time at the Café Opera.

The following morning, December 9, Bob and I were wide awake at 3:00am. The festivity, the twinkling lights outside, the cool air, swirling waters, and lingering effects of wine had converged to create a perfect storm of mental confusion. After consulting The Plan, I realized at last I had some time to take a real soak in our Swedish tub and happily submerged myself for the next two hours. This experience should be a requisite one for any guest in attendance during Nobel Week.

At noon, Peg Esber, Nat, Todd, and Todd's best friend, Paul Esber, were driven with us down slippery streets to the American ambassador's residence for another intimate Texas-style lunch for 150 or so, including the other American prizewinners and their families. Ambassador Lyndon L. Olson and Mrs. Kay Olson's palatial residence had huge fireplaces aglow and many large Christmas trees thematically decorated and saturated with silvery ornaments and lights which created glitz and glitter throughout the public rooms. The Olson family is of Swedish ancestry, but came from Waco, Texas, where everything is done BIG! So it was no surprise to be greeted by life-sized nutcrackers guarding the front door, smaller nutcrackers gracing the mantle of fireplaces, and nutcrackers marching with military precision around the top of the luncheon tables. Before lunch I took a self-guided tour with a brochure that catalogued the impressive paintings that hung throughout the house.

We expected a car-sized barbecue grill to be wheeled into the room any minute for grilling ribs and shucked corn, but instead we had a tamer lunch of grilled veal loin, tiger shrimp, potato pancake, vegetables, and a citrus fruit medley.

We were seated next to another American Ambassador, Reno Harnish, who was taking his family to be stationed in Cairo, Egypt, the following year. He kindly invited us all to come visit.

After the lovely lunch, we returned to the Grand for a ten-minute nap and a forty-minute clothing consultation. Next on the schedule was an appearance at the Nobel Foundation reception given in a stately building in Old Town that was originally the Swedish Stock Exchange. The impressive nutcracker-free room had golden gilt edging on the walls and crystal chandeliers as large as cars hanging from the vaulted ceiling. It was here that we were greeted by gracious foundation members and finally met up with some of our other guests and family members. Horst and Bob were able to introduce their mothers who enjoyed each other's company. Bob and I learned more about what happened the previous day, when our sons had apparently been taken with the other children and grandchildren of the laureates to a building downtown to be filmed for a children's science television program. They were given random materials and asked to construct something like a weather balloon that could be launched. Normally a simple task it would seem, but on that particular day, rain, sleet, and wind were raging, and our two California boys nearly froze to death. Lucky not to be launched themselves, they had actually enjoyed the novelty of the task. Just what else our sons did during their Stockholm visit remained a mystery until they confessed some years later. Their fondest memories were apparently of the colossal views out their windows, the fireworks over the harbor, their proximity to a real palace, and endless room service. Todd ordered cereal one morning and it arrived next to a dish of some white substance, which he supposed was sugar, and dumped it on his cornflakes. It turned out to be salt. So he reverted to burgers for breakfast. For entertainment in their room, the boys watched the one English speaking television channel available, which ran the movie *Armageddon*. They watched it so many times that Todd memorized all Bruce Willis' lines and grimaces.

Returning to the Grand late that afternoon, we tried to nap again and realized this experience in Stockholm was like having infants at home—no one ever sleeps but the infants. Our last stop of the day was at a restaurant Aja had booked for us called the Villa Kollhagen at Djurgirden for our friends and relatives—all thirty-one of us! The chef at the Kollhagen was featured on Swedish public television and his culinary arts were certainly magical. We all had the Christmas julbord, choosing from a smorgasbord of dozens of delicacies displayed on long tables spiked with vibrant red amaryllis and apple arrangements next to miniature Christmas trees studded with ornate straw ornaments. Tawny breads sculpted like reindeer pranced among the trees. Cinnamon, cloves, and allspice laced the air. It was a true holiday feast including real Swedish meatballs!

It was on this evening after we got back to the hotel, that my sister, Lynn, and my two brothers, Bill and Ned decided to explore the town. They walked out of the Grand and around a corner to a glass-fronted bar and dancing club called Cafet. As my brother Ned described it: "The line was a hundred people deep with paparazzi and tourists everywhere. The atmosphere was electric." Arriving at the door, the bouncer said: "Sorry. This is a private party." It was in fact an MTV premiere of Will Smith's latest movie *Enemy of the State*. My brother, not to be deterred, returned and offered $100 to the bouncer. He once again declined their admission.

They walked back to the Grand where the concierge greeted them: "How's your evening going?" he asked pleasantly.

"Not too well," they replied. "There was an MTV party around the corner at a club we really wanted to get into."

My sister remarked: "We're family of Bob Laughlin, one of the Nobel Prize winners."

The concierge said: "One moment, please." He left briefly, made a phone call and returned to the lobby. "It would be wonderful for you to join the party. Please go and give them your name."

So off they went back to the club, feeling triumphant.

As Ned later explained: "Inside, there had to be a thousand people from all over Europe fashion moguls, Hollywood stars and deal makers, local Swedish elite, and us. We were three suburban-looking, khaki-panted, cotton button-down shirt wearing middle-class Americans with our mouths wide open and clearly out of our depth. We found ourselves standing in silence most of the time, which we thought might make us look a bit cooler, but it had the effect of making us look even more suburban. Toward the end of the first hour, the host of the party, Jerry Bruckheimer (the famous Hollywood producer and director) was projected onto a thirty-foot wall screen to introduce the world premier of his newest movie. It was to be simulcast in three cities Stockholm, London, and Johannesburg. We were to be among the first people to take part in this type of event in celebration of the Nobel ceremony that year. After Bill and Lynn left, I stayed late into the night, paid the tab, and walked back to the hotel. The next day I slept late, and when I woke up, I tried on my white tie and tails while I had the television on. I saw a Nobel ceremony underway. Panicked, I squeezed into my clothes and ran downstairs, fearing I had missed THE EVENT. I was reassured that what I'd seen on the television was the Peace Prize being presented in Oslo, Norway. I had not missed the Nobel ceremony after all!"

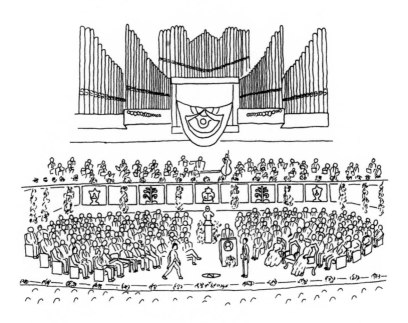

CHAPTER 7
The Nobel Ceremony

ECEMBER 10 BEGAN FOR US WITH A SERENE BREAK-
fast in the less formal of the two restaurants in the Grand,
which was full of other sleepy laureates all dishing up their
happy piles of pickled herring and smoked lox. A group of
blonde, pink-cheeked Lucia singers sang sweetly to us,
wearing white dresses with red sashes, one with a leafy
crown of candles, the others holding their candles in their
hands.

The day officially commenced with a visit to the
concert hall where the Nobel ceremony takes place. I wore
my new Swedish sweater like a security blanket as it began
to sink in just what it was we were doing there. As I
watched from the stage, Bob found his "N" (for Nobel)

mark on the center stage and rehearsed his bow to the king, to the academy members, and to the audience. The proper Nobel bow takes some practice, as one must bow low, pause, and then bow twice again while holding a large leather-bound watercolor and diploma. As the bows were always timed to music, mercilessly critiqued by the press and watched by millions of people, it was important to get it right. I was not as worried about Bob's bow as I was about his pants. Throughout his long lecturing career, he has had numerous pant-trauma events or PTEs that have proven challenging. For example, he gave a talk about a book he'd written at the Harvard Square Coop in which there was standing room only with people crowding him for autographs following his talk. At one point, he bent over to retrieve his computer's power cable and split his pants back to front. An Indian student cornered him for the next twenty minutes, during which time Bob considered what to do. With much diplomacy, he decided to hold his laptop at crotch level like a briefcase and smile a lot. I later surmised that Bob had remembered Miss Marple always carrying her purse in front of her on thoughtful walks through St. Mary Mead and decided if she could do it, so could he. After leaving the Coop, Bob had reluctantly boarded the brightly lit T subway to arrive at a distant parking lot, where he got into his rental car. He then drove two hours to Smith College in North Hampton, Massachusetts, stopping at a McDonald's at midnight (with even more leg ventilation) and was fortunate to escape arrest as a flasher. Hence, my apprehension about his bows.

After the laureates were shown their seat assignments onstage, entrances, and exits, Bob returned to the hotel for a BBC interview and to practice his bowing if he had time. Aja took me to my appointment with the hairdresser at Tre Klippare, where she also had her hair styled. We were the only ones in the salon and were treated like royalty with tea and chocolates to fortify us. After approximately two hours, my shoulder-length auburn hair was uplifted and transformed into an elegant French twist with lots of top curls and jeweled ornaments tucked into shiny folds of hair. I was unable to recognize myself. The inner and outer transformation that one experiences in Nobel Week is life-altering, not just for the laureates, but for all associated with them. As a spouse you know the many long hours that your partner has worked to get there, demonstrating the rare intellectual tenacity and integrity needed to discover truth in our natural world. The magnitude of the creative accomplishment and the way in which your support as a partner has played a hand is the meaningful part of the transformation. The prize is the vehicle of that realization.

I am pleased to report there was no significant apparel trauma that afternoon in our suite. We ordered club sandwiches from room service and checked on our boys next door. Then everyone in the hotel associated with Nobel Week started changing into formal evening attire.

Bob looked dashing in his white tie and tails. I wore a black velvet bodice with a red satin skirt, embossed with black velvet flowers, along with a red satin shawl that reversed to black velvet. Gloves were not required. For jewelry I wore an intricate, glittery silver floral necklace with matching earrings.

My parents were also preparing for the evening and, as my father later explained: "Precisely at 3:00pm, there was a knock on our doorvand the delivery of my rental stuff was concluded. My nervousness dropped. I had a scotch and soda from the in-room bar and took a shower under a flow of soothing Swedish minerals. The bath towels were massive and soft. I may even have hummed a snatch of *Ode to Joy* to dress. Everything went well until I put on the dress shirt and discovered the winged collar was choking me and wouldn't button with the stud. I called Carolyn who was finished dressing herself stunning as usual—and a panic attack took hold, since we were due down in the lobby at 3:30 for photographs and to join the other Nobel guests for the drive to the ceremony. As luck would come the tailor's store name was Hans Allde and the phone number was on the box of the dress suit. The phone call worked perfectly and in seconds—yes, the store was still open, and yes, they were very sorry and yes, they would deliver a dress shirt immediately. How? Heavy evening traffic, limos lined up outside the hotel, how big was my neck? I couldn't have eaten that much lobster! Would they get through a mobbed lobby, up several floors, and down a wide hall to our room? Would the elevator or stairs be quicker? Hundreds of men that night had probably rented evening clothes; we were going to meet the king and queen for God's sake! A corsage of orchids for Carolyn and one with a carnation and fern for me were at the ready."

Miraculously, Dad's new shirt arrived in time, but this incident was proof positive that my garment trauma was genetic.

We met with our entire family in the lobby of the Grand Hotel at around 3:30 and were taken to a private room to have a formal group photo taken. This photo remains one of my favorites, as everyone looks glamorous, a bit apprehensive, and so proud. At 3:50, we dispersed to various forms of transportation, all choreographed to perfection. Bob and I slid into the backseat of our limo. Pitch-black skies had descended on the city, but the white lights shimmered to illuminate the slick sidewalks. Our caravan of limos received a police escort through the streets of Stockholm, on which all traffic had come to a halt. Hundreds of people silently lined the streets, standing next to waist-high hills of snow that had been scooped out in the center and filled with flaming torches. The contrast of hot and cold, light and dark was startling. It was a primal sort of image that neither of us has ever forgotten. This path of light led us to an illuminated concert hall.

Our sons and extended family had been transported

behind us by bus and we reunited with them at the concert hall. Bob headed backstage looking a bit overwhelmed. Unfortunately, not all of us were able to be in the audience together. Relatives of the laureates were seated in a special viewing room with other families to watch the ceremony unfold. Bob's mother, my parents, Bob's three siblings, our two sons and I were escorted to the second row of seats inside the hall. The first several rows on the right of the orchestra area were reserved for the laureate families. Mrs. Pople, wife of the Chemistry winner, John Pople, sat directly behind me and I worried that my hair creation would obscure her view of the stage. To my left sat members of Swedish society with the ladies in lavish gowns and jewels, the men in tails with medals and sashes across their chest. There was a total of eighteen hundred people in the concert hall as Swedish television broadcast the event.

My immediate impression as I looked at the stage was my disbelief at the profusion of flowers. The program stated: *In homage to Alfred Nobel, who spent his final years in San Remo, Italy, and also died there on December 10, 1896, the City of San Remo sent the flowers to decorate the Concert Hall and the City Hall. In the Concert Hall, there were 8,500 carnations in different shades of orange and red-orange; 1,300 orange gladiolus blossoms; 1,000 yellow lilies; 200 orange gerbera; and 150 apricot amaryllis blossoms.* The flowers cascaded over the front of the stage and rose up to wrap around surrounding columns that were also an apricot color, and across the front of the balcony. Breathtaking. At precisely 4:30, the royal anthem called "Kungssangen" was played, followed by the *March in D Major* by Wolfgang Amadeus Mozart. The Royal Stockholm Philharmonic Orchestra was located in the balcony above the stage area. The chairs on the right of the stage were already occupied by members of the Royal Academy of Sciences, the Nobel Assembly, the Swedish Academy, and a representative of the Norwegian Nobel Committee. Any previous Nobel winners who had come to Stockholm for Nobel Week were seated on the left. Their Majesties King Gustaf and Queen Silvia of Sweden and Her Royal Highness the Duchess of Halland entered the hall to be seated, along with the nine 1998 laureates. Bob sat in the front row of the laureates' seats beside Horst Stormer and Dan Tsui. My mother later remarked that in that moment she felt overwhelmed by the beauty and grandeur of the occasion and seeing Bob join that assemblage of such brilliant and creative minds was an incomparable experience.

An opening speech was given by Professor Behgt Samuelsson who was chairman of the board of the Nobel Foundation. This speech was followed by the fourth movement from the Symphony No. 3 by Franz Schubert. Following this interlude and a short address, the Presentation of the three Nobel Prize winners for Physics were introduced by Professor Mats Johnson. Professor Dan Tsui, Professor Horst Stormer, and Professor Robert Laughlin who rose to receive their Nobel Prize gold medallions from the king. The Chemistry Prize followed to Professor John Pople, then the Prize in Physiology or Medicine to Professor Robert Furchgott, Professor Louis Ignarro, and Professor Ferid Murad. The Literature Prize was next and was given to Mr. Jose Saramago. And lastly, the Economics Prize was awarded to Professor Amartya Sen. An operatic solo was performed and musical interludes were played between the categories of prizes. The ceremony ended with the Swedish national anthem *Du gamia, Du fria*.

We remained standing while their Majesties the King and Queen of Sweden left the stage. We were all left emotionally drained not only by the ethereal singing of the operatic soloist Katarina Dalayman, but also by the sheer magnitude of the event. I had cried when Bob's name was announced and he spotted me as he bowed to the audience. As public as the gesture was, for us the intimacy of the moment was poignant indeed. I was, and remain, so very proud of him.

The newspapers ended up flunking Bob on his bow to the king, but gave him the highest marks for sheer charisma. Our sons had remained awestruck throughout the ceremony, but were delighted to go onstage following the presentations to have photographs taken and to hug their father.

CHAPTER 8
The Nobel Banquet

$\mathfrak{J}$UST WHEN YOU MIGHT HAVE THOUGHT THINGS couldn't possibly get more impressive, the evening continued with a banquet at the Stockholm City Hall. As 1,240 guests and students arrived, five hundred torches lit the way, illuminating the front façade of the City Hall. Bob and I went in through a side entrance in the southern portico of the City Hall that led into a private reception area for the laureates and for those guests who would be seated at the Table of Honour (Head Table). With sloshing champagne in hand, Ambassador Paulo Castilho of Portugal came forward to shake Bob's hand. Other ambassadors present included those from England, Norway, India, and Germany. There were also members of the Swedish government there. We formed a long line to greet the King and Queen of Sweden in the Prince's Gallery. Both were very kind, spoke perfect English, and welcomed us to the banquet. The queen wore an elegant apricot gown and a fabulous pearl necklace. Her earrings were studded with antique cameos and her grand tiara also held a number of gorgeous pearl-lined cameos that complemented her ornately beaded gown. Their Majesties both seemed quite calm and as my mother later commented: "You'd never have known from her demeanor that the ceremony had to be delayed because of a bomb scare outside Concert Hall after everyone else had entered."

After our introductions, we were assigned an escort and continued our procession line down the long Prince's Gallery in which the Nobel Prize diplomas and medallions were displayed in glass cases for viewing by other guests later in the evening. The leather-bound diplomas all had an original watercolor by a Swedish artist inside on the left cover of the folio and a hand-scripted page on the right side that contains the recipient's name and a short description of his or her work. At 7:00pm, we all crossed a balcony on which an organ and trumpets played a fanfare to announce the arrival of the laureates. Borrowing from the program's description: *The guests of honor processed down the grand stairway into the Blue Hall. The Blue Hall was decorated with four shades of roses. Fifteen florists participated in the floral decoration work. The entry procession of guests of honour was led by the Master of Ceremonies bearing a staff presented in the Nobel Jubilee year of 1991. Behind the master of ceremonies walked two female attendants, followed by His Majesty the King with Mrs. Pople, and Her Majesty the Queen with Professor Bengt Samuelsson, Chairman of the Board of the Nobel Foundation and host of the evening's banquet.*

I was escorted down the stairs by the Royal Marshall, (H.E. Riksmarskalken), Professor Gunnar Brodin, or as he described himself, His Majesty's right-hand man. Tall and steady, Professor Brodin did a wonderful job helping me navigate the slippery stairs in a long evening dress. The walls of the City Hall are red brick and had been transformed into explosions of color by projected beams of light. The bricks were originally to be painted a polished blue, hence the name Blue Hall. A sky with clouds was projected on the vast ceiling and the room resembled a single, rare, pulsating jewel as one descended the grand stairway. All of the guests at surrounding tables stood as we entered. They had entered a half hour earlier. Our family was located at the base of the stairway, looking radiant in their finery. We were only able to seat a small portion of our guests inside the banquet itself, so arranged for the rest of our party to attend an elegant dinner at Restaurant Leijontornet in Old Town, with our sons acting as hosts. I had decided to relinquish our sons' seats in order to seat two older members of our party. It is my fondest wish to return to Stockholm some year with our now-adult sons and have them attend the banquet with us.

The Table of Honour was oceanic in its proportions.

It was to seat ninety dinner guests and had been meticulously laid with Nobel dinnerware and table linen that had been created especially for the 1991 Jubilee. The glassware all had golden stems. The dinner plates were placed on gold and white chargers, with accents of yellow, blue, and green in the china itself. After the king was seated, we all sat admiring the exquisite floral arrangements. On the table were six tall obelisks of crystal with gilded porcelain bases from which flowed red and lavender roses. I am allergic to some roses and noticed that no arrangement was positioned directly in front of me. This attention to detail seems inconceivable still. Cascading from each of these bouquets were several strings of pretend dynamite made from pieces of bamboo in honor of Alfred Nobel. Also on the table were scattered numerous gold foil-wrapped coins of dark chocolate that were imitations of the Nobel medal. The ladies scooped many of these into their handbags to distribute to children and grandchildren later. Great care was taken to be certain we understood that this was not the king and queen's party, but the Nobelists' celebration. Shortly after our arrival, a toast to Alfred Nobel's memory was proposed by His Majesty. A four-minute photo opportunity followed the toast, with photographers swarming into the room.

In front of us lay a sea of cutlery and glassware. To my right was Laureate Lou Ignarro, who leaned over to me and said: "Can you believe this? It's unimaginable. If only my Italian father had lived long enough to see me

here tonight." The first course was served by a flotilla of
two hundred waiters in white who had descended the
grand staircase. They brought all three courses in this
manner, serving thousands of plates of food with military
precision along with numerous wines to thousands of
guests. The dinner menu read as follows:

Fonds D'artichaut marines et garnis aux crevettes,
aux ecrevisses et au fenouil

*

Supreme de poulet au thym et au poivre noir de Java
Ragout de legumes de saison et petit rouleau aux
champignons, Sauce a la crème de topinambours

*

Glace Nobel

Glace au chocolat blanc et sorbet aux mures sauvages

VINS

Champagne Pommery 1991

Brut Milieime

Corton Les Bressandes

Grand Cru 1989

Prince Florent de Merode Ladoix

Sandeman's Tawny Port
Aged 30 Years

Various menus are proposed and practiced
throughout the year in Stockholm, leading to a final
decision for the Nobel banquet. Somehow, the cold

courses for this dinner were served cold, the hot entries, served hot, all 1,240 of them! This was all the more amazing because the kitchen was on the floor above the banquet hall and the waiters had to bring the food down a long corridor, then down the Grand Staircase, with great precision. It is considered an honor to be chosen as a waiter at the banquet and they rehearse for months ahead of time.

Bob was seated across from me at the table next to Dominique Parchett. Before dessert was served, Katarina Dalayman again performed magnificently for an appreciative audience from the ostentatious staircase, singing arias at the top of the staircase in a billowy gown. With every five steps she descended, she changed not only her music, but shed an outer garment, to reveal a new costume beneath! Throughout the evening, the musical interludes were as follows, to quote the program:

To feature three composers with strongly contrasting temperaments, but all well versed in the dynamic interplay between words and tones. They take us on an impassioned journey between chill and fire, from the ice-cold yearning of first love and music that resound from Nordic nature, and finally upon black wings of a bird flying into a tense night where bloodstained intrigues are hatched.

After the coffee was served, the time had come for the toasts. There was fanfare as each designated laureate rose to walk to the raised podium. Bob was third in line to speak and he bashfully approached the podium in front of all those distinguished guests, knowing that millions of people were watching on television all over the world. We took a collective deep breath. His toast was delivered as follows:

Your Majesties, Your Royal Highness, Ladies and Gentlemen,

I have been elected by my co-winners in Physics to propose to you a version of Mark Twain's famous toast to babies, for babies are the first contact most of us have with real physics and thus an important component of the Nobel Prize experience. And of course we all have been babies. All of us remember the first day those bundles of joy arrived and the cold realization that there would be no academic freedom in THIS house for the next thousand million eternities, for it is an experimental fact that time runs slower in the vicinity of babies, especially at night. Who of us cannot remember learning the true meaning of eternity by spending his first night alone with an alert baby? And who would deny the iron certainty of Heisenberg's famous uncertainty principle, which states that two new parents cannot possibly get a good night's sleep simultaneously? As for black holes, I know they exist, for I have seen them and know them to be the most powerful things in this or any other universe. Yes, babies have a natural oneness with nature that makes them a vehicle for even the most abstruse concepts in our field. Take, for example, the destruction operator. It costs thousands of dollars and many years of study to fully educate an undergraduate about this concept. But just let a baby loose in a living room in which your stereo is down low within easy reach, and you will shortly understand with the most wonderful clarity what a destruction operator is.

My toast this evening is actually not to babies at all, but to the parents that put up with them, for they are making loving investments that may take a lifetime to bear fruit, and sometimes longer. My own father, for example, died a few weeks after I had done the first piece of work leading to the Nobel Prize this year, and so never knew what he had accomplished. There are other examples, but we need not belabor the point. Life goes on thanks to babies. So please join me in raising your glasses to parents, both present and not, who are the real heroes of events of this kind, for without them, there would be no life at all.

As I stated previously, there was not a dry eye in the house after the toast. Bob glanced tenderly at his own mother as he left the podium. Lou Ignarro, to my right, who had earlier mentioned his deceased father, could not speak for several minutes. This toast was in fact played over and over throughout the night on Swedish television. There has since been a worldwide traveling Nobel exhibition in which the laureates and their work are highlighted. In one section of the exhibit, there is a listening room in which toasts made by several former laureates can be heard. Bob's toast was selected as one of those toasts.

Nearly as impressive as Bob's toast was the dessert!

The lights were all dimmed in the Blue Hall. Down the long staircase, one step at a time, came the waiters in white, each holding a platter at shoulder height that looked like a salmon-colored dome of light. Inside the ring of light was a mound of ice cream surrounded by meringue. On top of each dome was a freestanding letter "N." A woven dome of white spun sugar encircled the ice cream. Masterful, magical, and unexpected, the creations took our breath away.

The banquet officially ended at 10:25pm, but the evening was actually still young. In the course of the dinner, I had lost my right shoe and spent a few frantic minutes at the end of the banquet looking for it! Looking south beneath the Table of Honor, I finally located it and slipped it on in time to proceed back up the Grand Staircase from the Blue Hall. Standing on the balcony, we were greeted by students from Swedish universities and colleges who were bearing the flags and standards of their student unions in order to pay homage to the laureates. We were then led into the Golden Hall where the ball was to be held. This room literally shimmers in gold from floor to ceiling. Its design was inspired by the Byzantine era. The nineteen million fragments of gold leaf that make up the mosaics on the walls were conceived by Einar Forseth (1892-1988). The Kustbandet Orchestra was playing as Bob and I joined the others and danced. We also visited again with the king and queen and greeted our personal guests who had been extended tickets to the ball and were thoroughly enjoying themselves. While I was chatting with them, Bob was brought into an adjacent room and interviewed on Swedish television about his remarks in his toast.

After that, we walked down the Prince's Gallery to look again at the watercolors and the diplomas of the other laureates. By then it was midnight and we found our driver and headed for the student Nightcap parties that were being held at the University of Stockholm School of Economics. Each year, the Nightcap is held at different university locations and the students expend superhuman energy transforming their buildings and residences into elaborate theme parties. The theme for this year was "The Festivities of the World." As we entered the first building of "party central" we officially passed into the inner sanctum of the Nightcap. Unfortunately, our sons were not yet eighteen and could not attend with us.

This was an enormous party. The featured celebrations included an Asian room with karaoke singing; a Middle Eastern room with baskets of sweet dates inside a Bedouin tent; a Swedish Christmas room with hot glogg; a Mardi Gras room featuring jazz, Cajun food, and beaded necklaces; a Swedish mid-summer solstice room with pickled herring and salmon; and a French wine festival room with select wine, crackers, and cheeses. There was a medieval feast with jugglers and jesters, a Valentine's Day room stuffed with red pillows and champagne flutes, an Inuit festival with furs and vodka, and a Rio carnival with sand, sombreros, ruffles and beer. There appeared to be cargo planes of food, tankards of drink, and dancing at every corner with hot music of every beat melting the frost off the windows. We enjoyed talking to many drowsy Swedish students who had been awake for several days creating this memorable Nightcap.

In this colossal Nordic Never Neverland of pandemonium, we managed to locate some of our guests and relatives. My siblings apparently had a great deal of fun. Lynn, Bill, and Ned had been generously transported to the Nightcap party in Bob and Anne Bass' limo. Lynn played with several colorful bead necklaces that she had picked up from the Mardi Gras floor. Ned reports that when Lynn joined them, Mrs. Bass asked her: "How did you get those?" Lynn, noticing that Anne Bass was wearing a sparkling necklace of her own, replied: "I got mine on the second floor. I must have missed the floor where they were passing those out!"

After a few beers, Ned tried to convince Mr. Bass "to fly us to Russia for dinner or dessert" in his private jet. Bob Bass said that the problem with that idea was that once the plane lands, the Russians are reluctant to give the plane back

Ned, who had begun to work in the mortgage business, asked Mr. Bass: "What types of businesses do you invest in?"

He responded: "Mostly prosaic businesses."

Ned asked him to describe a "mosaic business." Bob

politely corrected Ned and said: "Prosaic businesses." Ned, feeling mortally embarrassed, nodded his head in agreement, and made some gesture indicating that this sounded more logical. Having no idea what "prosaic" meant, Ned asked his parents at breakfast the next morning. "Once I got the information, I got back on the horse."

By 3:00am, I announced in my quiet way that if we didn't leave the Nightcap, I'd need to be carried back to the hotel on a stretcher. So, reluctantly, we left. Meanwhile, our sons and other teen guests had been taken on an extensive limo ride around Stockholm. Now that they are much older, I have no doubt that our sons would covet a trip to the Nightcap party!

CHAPTER 9
Dining with a King

Ｄ ECEMBER 11: DINNER AT THE PALACE. THE PROSPECT of having an intimate dinner with the royal Swedish family was certainly not an invitation to refuse, but I was petrified. That afternoon, Bob was taken to the Palace Library for an interview and had looked at some of the extensive array of book titles. He seemed relaxed and eager to dine in a palace with a king. He had even taken the time to read the biographical material we were given about Their Majesties King Gustaf and the Queen.

While Bob was enjoying himself at the palace, I was again at the hairdresser. Nervous beyond reason, I felt some level of reassurance when my hair stylist commented that she traveled to Paris twice a year to learn the newest styling techniques and fashions. Thankful that she was not going to rely on my hair suggestions, I let her loose. I described my gold-and-turquoise Thai silk jacket to be worn with a satin gown. I described my brand new gold slippers. I talked non-stop about the ceremony and banquet and the Nightcap. As I talked, she created. I ended up with a braided hairpiece that added volume, and a wildly original sculpture of twists and turns studded with real leaves, rhinestones, and maybe even a fake bird, I wasn't certain. I thought I looked a little like a portrait on a china platter of Marie Antoinette in the French court. Again, it was a transformation that left me speechless.

Returning to the Grand, I attempted to nap, which is not an easy maneuver with forty-five pounds of stiffened hair barbed with 702 bobby pins. Bathing was also out of the question, so I sat bolt upright like a China doll until it was time to dress and head for the lobby. Our *Nobil-12* limo transported us to the palace and we walked up a royal-blue-carpeted stairway into the Reception Hall. It had an expansive polished inlaid wood floor. Along the walls were glass cases filled with antique silver and Sevres china. The vaulted ceiling had an immense mural delicately depicting heaven and the angels, with gold angels hugging the corners. There were seemingly dozens of brass clocks that chimed along the length of the hallway and into the room where we were to have dinner.

The Nobel laureates and their spouses formed a receiving line so that the royals and other guests could meet us as they arrived. The master of ceremonies approached, banging his staff against the floor to command attention. He headed briskly in my direction and leaned over to whisper in my ear. It seemed I was to be escorted into dinner by Carl XVI Gustaf Folke Hubertus, King of Sweden, and I was to move as quickly as humanly possible across the polished floor to the front of the line. Immediately! Gathering up my dress, I made a valiant attempt to walk like a lady to the other end of the room, which was about the length of a Disneyland parking lot. My gold slippers were like freshly sharpened ice skates. I literally slid sideways into the King and said: "I think I should tell you my shoes are a bit slippery." As the words popped out of my mouth, my renegade right foot shot out in front of me. Owing any sense of balance I have to years of near-death skiing mishaps, I regained my composure quickly without actually landing on top of the king, who too, was clearly dazzled by my performance. He took my left arm securely and led us into the Gallery Room. In it was a table that I approximated to be one-quarter-mile long. Laureate Doug Osheroff later corrected me: "If we sat three feet apart, the table was 232 feet long, seating 155." In any case, I literally could not see to the other end. I thanked my stars that His Majesty knew where to sit. We stopped halfway down the table at his royal-blue chair. I was seated to his right and Literature Nobel Laureate Jose Saramago was to my right. Across from me was Laureate Pople, and to his right was the glamorous and multi-lingual Queen Silvia. Bob was across the table to the left six seats down and I was barely able to see him. So there I sat, in my turquoise silk gown with my Marie Antoinette hairstyle, next to a king in his own palace and definitely not in Kansas anymore.

At the table, we had a manservant standing at attention behind each of throughout the dinner. The servants pulled out and pushed in chairs and tended to all errant napkins and empty wine goblets. The centerpieces on the table were enormous silver vases, two or three feet high, with floral arrangements cascading from the top like soundless fireworks. At the base of each vase were sculptures of nude women dancing in a circle beneath the bower of flowers. There were also on either side, two silver nude women holding two silver bowls, filled with salt and pepper. Whenever His Majesty wished to see the queen, he had to look through the silver dancing legs in front of us. Resisting the temptation to comment upon the preponderance of naked women on top of the table, I turned to my right and introduced myself to Jose Saramago. Saramago, unfortunately, spoke no English, and I spoke no Portugese, so it was a short conversation. I did my best to pantomime by pointing out the beautiful flowers. He nodded and we both nervously scanned the menus on our plates.

When I was young, my father was a college president and my parents did a great deal of entertaining, so the words "canapés" and "hors d'oeuvres" and "crudités" were in my vocabulary by age nine. By age ten, I was pouring mixed berries spiced with warm cognac into the centers of stiff meringue shells and serving them (on a person's left side) at formal dinner parties. I knew about the dessert fork and spoon at the top of a place setting. I knew that the wine goblet went to the right of the larger water goblet. I knew the red wine goblet was usually taller than the white wine goblet. In front of me now, however, there was so much extra cutlery and glassware that I grew a bit faint. I remembered my mother's dictum: "Always start from the outside of the silverware and work your way in toward the plate." That sage advice saved me from profound embarrassment throughout the evening. Another strategy would have been to watch the forks and knives used by Queen Silvia, if I'd been lucky enough to be seated next to her. I did wish, though, that an explanation of glassware and cutlery protocol had been part of our initiation talk at the Grand. In retrospect, part of the fun at these formal gatherings is to see just what it is that people do with their place setting "tools." I am sure it crossed the mind of some flustered guests to slip a spoon or fork into a pocket for a memento, rather than try to ferret out its actual purpose for being on the table.

I had stalled long enough looking at the menu and the tableware, so decided to plunge into a conversation with King Gustaf. I had not received any guidance about what to talk about and what to avoid, so I did my best to pretend he wasn't a king at all and jumped in with the question: "Do you have ghosts in this palace?" That caught him a bit off-guard, as had my hair, but he answered with enthusiasm: "Of course! At the end of this table in that window down there, I have had servants stop and see a woman in the glass. These are reliable people, you understand. There was a woman who died in this room and her portrait is here somewhere."

He continued to explain that he was having excavations done in the cellar and finding the 1,400-year-old walls of a former palace. He commented: "I've lived here all my life, but you'd never catch me in the cellar at night!"

As the first course arrived, His Majesty told me he had just attended a moving performance given by retarded and handicapped children at an award-winning school in Sweden in which the retarded children acted out what it would feel like to be physically handicapped and the handicapped pretended to be retarded. He said he hadn't known whether to laugh or cry, it was so moving. I told him about my brother with Down's syndrome and his special challenges. I added that I had decided to work with special-needs children because of my brother. I knew that some dyslexia was in His Majesty's family and we briefly discussed how traumatic that disability can be for individuals.

Trying to lighten things up a bit and since it was December, I commented that Bob enjoys choosing and cutting down a tree for us at Christmas with our sons. The king replied: "It is illegal in Sweden to cut your own. If we allowed it, there would be millions of trees gone. I, however, decreed that I can cut down my own tree and go with my children to do it." Wouldn't it be nice to decree something, I thought to myself.

The venison was served. I remarked that I had never eaten venison. *Was I actually eating reindeer and would it be rude to ask?* The king, perhaps sensing my apprehension, offered: "I have my own hunting area. Venison is the traditional meat served at this meal. I am supposed to have shot it myself. I didn't, but we can pretend I did." I began to cut my venison with the special venison knife and as I did, the three hundred-year-old plate rotated a half turn on top of the silver charger plate below it. It was as though unseen hands were spinning it in front of me. There was no direction or pressure that could be exerted to stop the plate from spinning. His Majesty noticed and said: "I see you have one of the traveling plates. I try to avoid them myself. You see, they are so old and uneven on the bottom that they travel." Finishing this course with some degree of motion sickness, I did not have the courage to clarify the reindeer question and decided that some things are better left to the imagination.

Relieved, I watched as a whimsical dessert arrived and was set before us. There was a mound of custard in the middle of the plate, with a long chocolate straw lying on top of it. On the north of the plate was a spoon made out of cookie dough that had a dollop of sorbet inside it. I remarked that the dessert looked like a physics experiment. (By now, I had sampled numerous beverages, and Bob was not within hearing distance anyway.) His Majesty joined in and said: "Yes, rather like Uri Geller. Shall we try to bend the spoon?" As he picked up the spoon, it snapped into two pieces, and his hand dropped and hit the chocolate straw, which then catapulted magnificently into the air. After what seemed like an entire minute of flight, the chocolate straw landed in the lap of Lou Ignarro's wife, Sharon, who was seated to the king's left. At least, we both thought it had landed squarely in her lap. King Gustav turned to me and asked: "Oh my! Do you think I ought to retrieve it?"

I responded: "I think not."

I cannot recall what he did next, as I was convulsing with laughter. I did smother my mouth with a napkin to try to regain some degree of composure to befit my regal hairstyle. Each time His Majesty looked in the direction of his dessert plate, he laughed, too. Luckily this incident marked the end of the meal and I was then perfectly free to make a fool of myself again out on the parquet floor. As I followed the king out of the room, he turned and graciously extended his arm again as we left the dining room. We navigated the slippery floor without incident and he thoughtfully deposited me on an area of thick rug near a fireplace as I recall. Guests were then offered cigars from a humidor the size of a small suitcase. Coffee and cognac were also served. His Majesty had an ingenious way of holding a coffee cup and saucer while balancing a snifter of cognac above his wrist, freeing the opposite hand for his cigar. Bob came over to join us and I was unsure whether he had witnessed the flying chocolate straw incident or not, but it made a great story to tell him later. At the stroke of eleven, the queen's Ladies of the Court (recognizable by their black dresses with crisscrossed puffy golden sleeves) tiptoed around to gather people off their comfortable couches so that they could stand at attention as the king and queen departed.

We all proceeded back down the blue staircase and into a coat room in which there was a seven-foot mirror with a silver comb and brush lying on a table next to it for the use of the guests. The doorman called for our limo, which was parked in a huge courtyard, using his microphone to call out our number. Of the hundreds of cars, we were able to identify our own *Nobil-12* and walked out into the cold night air after a thoroughly delightful dinner with a king.

CHAPTER 10
The Nobel Foundation

THE DAY AFTER DINNER AT THE PALACE, BOB HAD a 9:30am appointment at the Nobel Foundation headquarters to officially receive his prize money. Inside the foundation building, we walked down a narrow passageway in which lighting focused on sculpted busts and paintings of Alfred Nobel. On the second floor, we viewed paintings submitted over the years by Swedish artists for consideration to go alongside the diplomas that the winners receive. One of the paintings was especially vivid: it was of a jagged barbed-wire enclosure with a bird caught inside. This was a painting submitted for Alexander Solzhenytzen but had been deemed too graphic.

Entering a darkened room, we were seated at a huge wooden table. On the table was each of the recipient's

medals and opened Diplomas. Bob was asked to stand at
the head of the table and to sign his name twice for a
special album of laureate photos. We were next led into
an adjacent room, where we discussed our preference for
the transfer of the funds. Bob said magnanimously: "I
think we'll actually leave it in Stockholm to pay our hotel
bills, then we can transfer it to my account." Bob was to
receive one third of the million dollar prize, to be shared
with Horst Stormer and Dan Tsui.

Following this exchange, we were driven to the
Institute of Technology. It was here that a panel discussion
about the "Future of Science" was staged, with Horst
Stormer, John Pople, and Bob debating the issue. Stig
Hagstrom was the moderator. Dr. Pople said he had
attended and now worked in the same institution and that
it had served him well. He also insisted that students
should not be bombarded with academic choices, but
should be allowed to focus on one and only one problem
that interests them. Bob disagreed, stating that he felt
students should expose themselves to a wide range of
studies in order to have the flexibility to change careers
later in life. He felt that students need the flexibility to
make changes as the venues of science change. The debate
continued for some time and as we left, the institute
presented us with lovely flowers. A luncheon followed at
which we were serenaded by Swedish students.

An hour or so later, we were driven to a reception
given by the Stanford Alumni Association of Stockholm.

We were generously presented with a Swedish wooden horse known as a Dalecarlian Horse. Its design dates back to the 1800's and is painted orange with a garland of painted flowers around the neck and a congratulatory plaque on the side of the horse.

Back at the Grand, we changed clothes and collected our sons, assorted relatives and Aja to have dinner at the home of Dr. Lis Granlund, who had been introduced to my parents by a mutual friend. Dr. Granlund was the retired curator of The Royal Palace and Drottningholm Palace. She had also been president of the International Council of Museums Decorative Arts Committee. She is a china and silver expert of some considerable reputation. Dr. Granlund had been traveling extensively in her retirement, staying at the residences of ambassadors she had met over the years. Her small apartment was packed from floor to ceiling with objects that chronicled her life with her husband of fifty years. One room held his nautical collection, with paintings of seascapes, ships in bottles, and barometers, etc. In another room was a silver tray with glasses and a sherry decanter set next to family portraits. Her salon was divided into two sitting areas with plush couches, needlepoint pillows, a tall clock dating back to the 1700s, a very old piano, and a sling-seat chair. Interestingly, the apartment was filled with abstract paintings on the walls. Lis had nestled a display of her antique Christmas dolls beneath a chandelier full of lit candles in the center of the room. On another table she had positioned an elegant fruit platter with grapes and pears next to a silver tray of demitasse cups. Small tables were festooned with embroidered Christmas cloths and plates filled with diamond-shaped Danish cookies. She used every inch of her space beautifully.

Twelve of us had assembled in her apartment and first were served a glass of sherry and pretzels. Then Lis seated us snugly at her table, pulling out each course of the dinner from a room directly behind her like a magician. The plates she used were one hundred years old and wobbled only slightly. "Aren't I terribly courageous to try to seat all twelve of you here?" she mused. We drank from delicately etched wine goblets and sampled salmon and bread, followed by pate, small pickles, and a Christmas salad of cold rice and fruit. These courses were followed by a lettuce and tomato salad and then by a huge platter of soft cheeses with radishes. Apparently about then, Todd whispered to my sister: "I don't know what all this is, but I think it's safe to eat anything green." This lovely dinner ended with a chocolate mousse cake and coffee in her salon. Dr. Granlund's husband had died the year before and she confessed that this was the first entertaining she had done since his death. We all expressed our thanks to her for such a marvelous evening together.

CHAPTER 11
The Supreme Order of the Ever Smiling and Jumping Green Frog Initiation Ceremony

NOVEMBER 12, WE HAD RECEIVED AN INVITATION from the Student Union at Stockholm University to attend a "celebration of Lucia and the dubbing of new members of The Supreme Order of the Ever Smiling and Jumping Green Frog on the evening of December 13." The invitation went on: "This order has existed since 1917 and the first Nobel Prize laureate was dubbed in 1936." I must confess this all sounded highly suspicious. I had been tempted to toss the letter out, but then heard Bob nearly suffocating with laughter as he re-read the invitation aloud in the next room. We had it on good authority from a 1997 Laureate, Bill Phillips, that this initiation ceremony was not

to be missed. He didn't go into great detail, but urged us
to attend. The invitation looked quite official, signed by
"the President and the Superior Steward," so we had asked
Aja to put the evening on our agenda. She later informed
us we would be able to take all twenty-one of our
remaining guests with us! Little did Bob and I realize that
this was to be the longest day of our natural married life.

The day began with a visit in the Grand's less formal
restaurant to have breakfast with the Wahlberg family.
Their son, Niklas, had been in my second grade class in
Palo Alto the previous year. We had a chance to get reac-
quainted while the Lucia singers serenaded us at our table.
Niklas was not speaking English too readily, but his
parents and I chatted. Bob found us and gave Niklas a
Nobel poster, which he then signed for him. This was the
first former student I had visited internationally and it was
fun to reconnect with his family. His sister presented me
with small St. Lucia dolls that she had crafted. I still get
them out every December, along with our Swedish
wooden horse collection and blue candlesticks.

At 9:00am, we were driven out to Uppsala University. This is an impressive campus of buildings with
rotunda ceilings and marble pillars. First there was a coffee
reception in the board room next to portraits of kings and
queens hanging beneath the lofty vaulted ceilings. Bob was
taken to give a lecture and I was taken on a walking tour.
The first stop was at a cupola designed by Olof Rudbeck

in which there is an anatomy lecture hall. There are slanted windows along the top with room beneath for two hundred spectators to stand and gaze down on the autopsy arena. Though the room was the first of its kind to be built, it was only used ten times. Autopsies would last for twenty-four to thirty-six hours or until the stench was unbearable. The bodies used belonged to criminals, some reportedly executed prematurely. Afterward, the remains were carried out respectfully in a solemn procession and then buried.

We were also shown a magnificent traveling case called the Augsburg Cabinet that was first presented to the Swedish king four hundred years ago. It is essentially a case in which there is room for all manner of personal effects. The lid of the case is most peculiar, with rare minerals such as amethyst, malachite, rock crystal, and black coral, all nestled together with seashells like an organic eruption. On the very top is perched a golden goblet with representations of the sea god, Neptune, and of Venus, the goddess of love. The case also contains hundreds of biblically-themed inlays, as well as multiple mirrors and tables that fold in and out. We continued through to a room that held an early physics apparatus, then walked to see a magnificent cathedral.

Laureates and guests were then reunited at the castle for lunch. This particular luncheon was held in a huge hall with a half-dozen stunning pastoral tapestries on the walls that depict royal outings. This hall had been the scene of many royal parties in the past, including one during which a barn full of peat caught fire a mile away that the royals surmised had been created for their amusement. One of the most memorable events of our entire trip to Sweden was when we were lucky enough to be serenaded at lunch by the Student Choral Ensemble. Their Christmas music CD was later mailed to us. It is a recording by an exceptional group of singers which we continue to treasure.

After leaving the luncheon, I was deposited once again at the hair salon. As I walked in, the comforting voice of Tony Bennett was smoothly singing "White Christmas" somewhere in the background. My hair arrangement on this last salon visit was probably my favorite styling of the week: several French twists with strings of pearls looping up and around to the top, creating a slight tiara effect. I had brought a pearl necklace and earrings to go with my gold-and-white silk jacket and evening gown. Returning to our room, I changed into my gown along with my infamous gold slippers, helped Bob and the boys with their evening clothes, and headed off at 6:30pm for the party at the Student Union of Stockholm University. We were fortunate to be able to take so many relatives and guests with us to this event, and even though none of us knew quite what to expect, we had been promised a memorable evening. Our attaché, Aja, whom we had by now virtually adopted as a daughter, had quietly removed her emerald gown from the trunk of our limo, found a changing room, and emerged looking like royalty. She had been such a help to us all week but that night proved to be perhaps her hardest assignment yet. She was in charge of "the kids' table," situated at the rear of the dining room. We had invited all our teens and pre-teen guests, including our sons, the lovely Libby daughters—Elizabeth, Samantha, and Alexandra, cousin Bob Martin, and Todd's friend Paul Esber, all looked dashing in their formal garb. Heaven only knows what they consumed that evening, given the eight drinking goblets that were positioned at each table setting. I do know there was Christmas punch and snuff delivered to the table. The teens also amused themselves by making and shooting paper airplanes above the candle flames. They also got a whiff of schnapps and learned to skoal. Beyond that, I would rather not know.

To faithfully set the stage for this party, picture the serious formality of the Nobel ceremony and banquet, and then imagine an evening that is its polar opposite. The evening with the Student Union is their "version" of a ceremony and dinner befitting a Nobel laureate who is to be initiated into their "club." The raucous evening had moments that bore some resemblance to the actual ceremony and banquet, such as the delicious food, which was served by fleeting barefooted "nymphs" who were female students. The menu was as follows:

Tartar from Norrland on arctic style bread
Fillet of reindeer and västerbotten

cheese pancakes with layers of
Chanterelles and sweetbread of veal with red currant sauce
Citrus parfait with strawberries and diced oranges
in a basket of honey

The dessert was, of course designed to mimic the spun sugar dessert at the banquet. On this occasion, however, it was presented with the aid of what appeared to be rescue helicopter searchlights that illuminated each platter and its server, while blinding the rest of us.

I was unable to identify what was in the arctic bread, and I thought it probably best not to ask. The courses were all brought forth with great gusto and fanfare, down a ramp of stairs. The beverages were poured like waterfalls spewing into so many glasses it was difficult to find dry dock for the dinner plates. The evening was punctuated by a series of skits and twenty-four drinking songs. My seat was at the end of a long table right next to the stage. This proximity to the stage meant that I had to stay awake, clap, sip, sing, skoal, and feast for what seemed like the next twenty-four days. One humorous skit that I do recall early in the evening involved a "Swedish plumber" who was asked to replace a pipe beneath the feet of the president of Stockholm University. That was the skit. I rolled my eyes in the direction of my father who I noticed was seldom in his chair, but was instead often being led by a gorgeous female student in the direction of the men's bathroom. Bear in mind that at the City Hall, we'd had to sit for three and a half hours dining without using a restroom. Here, I thought, how nice, Dad has an escort to the bathroom at frequent intervals. But at this banquet, people were being herded into the lavatories at alarming rates and I got suspicious. Leaning over to the handsome male student to my left, I asked about the people swarming into the restrooms. "Would you like to go too?" he asked enthusiastically.

When he saw my expression, he quickly added: "There is whiskey in the lavatories. Lots of pure whiskey." I declined politely and shoved a ball of arctic bread into my mouth.

I have neglected to mention that with most, if not all, of the drinking songs, one was required by protocol to stand on one's chair while singing and skoaling. The athleticism required for this undertaking was considerable, especially when emeshed in an evening gown, a full-length slip, slippery shoes, and waist-high pantyhose. Each time I ascended my chair to skoal, my pantyhose inched down my legs about three inches. After about eight songs (do the math) it was all over. The pantyhose had handcuffed my ankles together. I leaned forward to one of our Nobel hostesses, Christina Tilfors, to inquire how much longer the evening would continue. She assured me there was much more merriment to come. I asked if there were any way I could leave the merriment sooner. Hearing the panic in my voice, Ms. Tilfors promised she would do her best to hurry things along. Still landlocked in my pantyhose, I was leaning into the direction of my feet when the nice hazel-eyed young man next to me asked me to "dawnce."

"Maybe later," I responded spitting pearls out of my mouth as I sat up.

As the skits continued, I focused on extricating myself from my stockings and thought of a plan. With the ninth drinking song and ninth chair ascension, I slid my purse off my lap, leaned forward with my head beneath the tablecloth, disengaged my feet from the vice-like grip of the pantyhose, and stuffed the sweaty hosiery into my purse. Sitting up, however, I snagged a string of pearls from my hairdo on the tablecloth, nearly ending up with all the remaining plates, goblets, and arctic bread down the back of my dress. The cloth and I were like silent submarines colliding under a sea of white linen. Eventually I disengaged myself, felt the blood return to my lower half, and accepted an offer to dawnce leaving my odiferous purse beneath my chair. In the meantime, Bob, Horst, and John Pople were having problems of their own. They had been invited onto the stage and, while standing next to a huge frog mascot, were required to hop all over the stage as the first part of their "initiation." The men, after successfully hopping (better than bowing I suppose) received Order of the Frog necklaces. All three were then ceremoniously "beheaded" in a ritual that seemed to make sense to everyone except Bob and Horst who looked around nervously for what was next.

Expecting sautéed frog's legs, we got Horst. He gave a sobering talk (which he had managed to write in a day and a half while riding in limos and waiting for lunch at the castle). He mentioned people and inventions that should have received Nobel Prizes, but who had been overlooked. These inventions included paper clips, safety pins, and the rear-view mirror "so that women could put on make-up and pass cars simultaneously." As he continued, he invented physics names for these inventors, which of course only one-tenth of the audience understood. He ended by alluding to the four buckets of water that had to accompany Saint Lucia wherever she went in case she caught fire. Horst gave this speech with glasses perched on his nose in the most austere manner possible, which made his delivery stately and riotous simultaneously.

By midnight, my goose was cooked. I flagged down Bob and Aja and after several false starts, we collected our sons and located our driver. The evening for the other guests was to continue with dancing into the night in the next room, but we returned to the Grand. We hugged the boys good-bye because they were heading for California the next day without us. I spent the next hour and a half packing our suitcases so that Bob and I would be ready to be in the lobby by 7:30am. There was no adjective to describe the exhaustion we felt. The only words I managed to write in my journal (at 2:00am) were: "Next time wear tennis shoes and socks with gown."

CHAPTER 12
The Scandinavian Lecture Tour

𝕴T IS AN EXPECTATION THAT AFTER THE NOBEL PRIZES are given that the laureates will travel throughout Scandinavia to give a series of lectures to students and the general public. So, on December 14, Horst, Dominique, Bob, and I stumbled sleep-deprived into the lobby of the Grand and shuffled out to our limos for the ride to the airport. A taxi followed behind with the luggage. After a sad farewell to our fabulous attachés, we were whisked down a VIP tunnel and shown to our seats on the plane. The pilot politely introduced himself and promised a pinch of turbulence. Still experiencing severe turbulence from the night before, we all moaned and closed our eyes to try to sleep for an hour before arriving in Gottenburg. Once in Gottenburg, we were driven to our hotel, but Bob and

Horst had to go give a lecture. Dominique and I collapsed.

In the early evening, we were picked up for a reception at Chalmers University in a lovely old house with cobalt windowpanes in its front door. We had an informal stand-up dinner of caviar, salmon, shrimp, and chocolate mint cookies. Horst and Bob sat quietly, barely able to chew. As is the custom, we were once again serenaded by a choir, this time all male. The singers wore traditional sailors' caps with tassels dangling down one or both sides, giving them a slightly waterlogged appearance. Waterlogged or not, their four-part harmony was astonishing. For one of the songs, the students cuddled up to Dominique and me to sing a wooing love song called "Madeline." Horst and Bob were alert enough to demand a quick translation before the students departed. It was a relaxed evening and we were pleased to see friends Steve Girvin and Bert Halperin join the festivities.

After a night of normal sleep, we four set out, feeling newly resurrected, to Lund, a town three hours away at the southern tip of Sweden. An intrepid taxi followed our every move, filled with our bulging suitcases stuffed with clothing essentials. Enjoying one another's company on the drive, we watched a hazy golden sun graze the hillsides. It was the first sunshine we had seen in several days. The sun itself never seemed to be much higher than the horizon as it skimmed across the barren fields like a polished stone. Once in Lund, we stayed at a lovely place called Hotel Ideon Gasteri. It was run by a friendly

Swedish couple who host the prizewinners each year as they make their lecture rounds. The owners confessed that they travel to New York City for Christmas and to Key West for New Year's eve, which we all found quite amusing!

A formal dinner was given at the Bishop's House. The Lund University physics faculty was in attendance and I was thrilled that there was no chair climbing requirement. The university had purchased the Bishop's House and filled it with its own art collection. In the past, a Bishop had earned seventy times the salary of an average citizen, and no expense was spared in his residence. The rooms were stunning with red satin couches and small red poinsettias positioned throughout.

The next day, Dominique and I were escorted to explore downtown Lund. Lund is a lovely thousand-year-old town which meanders down cobblestone pathways and has numerous small shops, some dating back to the 1600s. The church in the center of town is magnificent. It is sparingly decorated, except for ornate wooden carvings on the pews with biblical scenes. It has a crypt with iron bars over it in which, legend has it, there lives a giant capable of shaking down the pillars of religion if released. There is also a huge clock with carved figures that emerge on top at noon in a circular dance. Lund is a peaceful town, built on a distinctly human scale with fragrant flower carts on each corner. Having been unable to find time to do much shopping in Stockholm, Dominique and

I located an Orrefors store. Orrefors furnishes all the crystal goblets for the Nobel banquet tables. As we entered the store, there were other customers looking around. I thought I might want to purchase some of the wine goblets. Dominique and I debated whether we ought to identify ourselves to the clerks as wives of 1998 laureates. One of us eventually spilled the beans, and the store was locked shut. We suddenly had the entire staff at our disposal and I decided I had to buy something. So I ended up getting ten red and ten white wine goblets—all gold-stemmed, along with a set of five-inch-high champagne glasses. These were immediately bubble-wrapped, boxed, and labeled on the spot to be shipped to California. They arrived at our house the same day we did and have been joyfully used for special family celebrations many times in the last ten years.

Early in the evening of December 16, Dominique, Horst, Bob and I took a hydroplane (in which we were the only passengers) to Copenhagen and were met there by a Danish professor, Poul Erik Lindelof. We were taken to the Palace Hotel, which faces the welcoming twinkling lights of Tivoli Gardens. This is a magnificent hotel that sits within walking distance of the main shopping district of Copenhagen. Our room was the "King Frederick" room and there was a large photo of the king on the wall as a reminder. The walls of the room were a shiny dark mahogany. There were "his and hers" bathrooms, a separate dressing room with floor-to-ceiling shelving, huge windows overlooking the central square, and a royal valance treatment over the bed. Ordinarily we would have been stunned, but as we were still pretending we were Marie Antoinette and Luciano Pavarotti, we sat down and thoroughly enjoyed the opulence of the accommodations. Plucking plums from our fruit platter, Bob decided maybe he could get used to this lecture business after all.

That evening, we attended a mostly male dinner at an elegant French restaurant called Philippe with the gracious Danish professors who were our hosts. Physics talk was rampant that evening, but special culinary mention belongs to the dessert, which was a chestnut-apple cake about the size of a cupcake that was served with a delicious cinnamon ice cream with sautéed apples and plums. This simple dessert was, in my opinion, the Iron Chef winning dessert of our visit.

The next morning, Bob and Horst gave public lectures at a symposium and then went on to visit the Orsted Laboratory. Meanwhile, Dominique introduced me to Copenhagen. There is a sophisticated ambience to this city with its posh pastry cafes, formidable blending of old and new architecture, decorative holiday windows, and high-end stores galore and wide pedestrian sidewalks that invite conversation. Dominique had visited the city several times and asked what I wanted to do. With my credit card still smoking, I suggested we window shop at the Royal Copenhagen, one of the premier china and crystal stores in Scandinavia. Visualize a half-dozen floors decorated with startlingly creative table settings—most of which incorporated natural items like bark, moss, tree branches, and pinecones. For instance, there was a pure white table that had loops of branches, pinecones, and ivy circling around the center of the table like race tracks beneath a chandelier cascading with ivy. On another table sat some tree fungus that had been rolled into large balls and now glittered in candlelight. Next to it were chairs cushioned with green moss. There were sheer white linens draped over tables with great white lilies on them mixed with glass ornaments and strands of gleaming silver threads. In another room were German star cookies with white icing placed next to chairs draped with sheer fabrics scalloped around the edges. There were tables full of moss gathered around blue-and-white china and candelabras sprouting white branches carrying hundreds of small golden birds. Glittering cones were mixed with fruits rolled in crystal sugar and piled into cut-crystal pedestal bowls. We also spotted china that was nearly identical to the settings we had seen at the palace dinner in Stockholm, perhaps replicas of the older patterns. These items were so expensive, however, that a special appointment was required just to inquire as to the price—and no doubt a dress code was part of the deal.

That evening we were driven to the airport to take a flight to Helsinki. The flight was delayed and Horst and Bob had grown tired and a bit punchy. By that I mean that the slight boarding delay had tipped the scales of sanity and they were both very close to losing their minds. Luckily we seemed to be the only ones waiting at the gate. Horst had been grieving the loss of his attaché for several days and occupied his time at the airport by riding an empty luggage cart up and down the hallway at highway speeds, calling out for his attaché. Bob, meanwhile, sat intently at his laptop laughing so hard tears were rolling down his cheeks as he typed. Dominique and I looked at one another with trepidation. This was going to be a long night. Eventually we all became curious and hovered over Bob as he typed. We discovered that he couldn't get the St. Lucia song out of his head and had composed anywhere from ten to thirty new verses of the song on his laptop. Belting out verses at the top of his voice, his new lyrics were hilarious, half of them needing serious spousal censorship. But on he sang. Boarding the plane, he sang. Landing in Helsinki, he sang. And yes, we still sing a few verses every year in St. Lucia's honor in the privacy of our own home.

We were an hour and a half late arriving at the Helsinki airport, which is arguably one of the most lavishly appointed airports in the world. It is dazzling—displaying elegant perfumes, expensive furs and crystal glassware, and full-sized reindeer rugs, under thousands of glittering ornaments suspended from a birch ceiling. Since we were late arriving, we were taken directly to the university, where high school students had assembled to ask Horst and Bob questions. I crossed my fingers that no one would mention St. Lucia. Most of the questions were quite personal and perceptive, such as: "How do you balance your home life with your professional one?" I think Bob answered true to form: "I don't. My wife Anita does that."

Afterward, we ate fish cakes in the hallway while Horst volunteered to be hooked up to a machine that measures brain function. We all thought this was very funny, as I myself doubted he still had any neurons firing after the week we'd all had in Stockholm. This test was to take place at the famous low-temperature physics laboratory in Espoo, Finland. It was a laboratory started by Ollie Lounismaa, a well-known and highly respected Finnish scientist. His heir apparent was Professor Mikko Paalanen, an old Bell Labs colleague of Bob's and a good friend. Mikko loved the idea of making Horst a sacrificial guinea pig. In 1985, they had apparently taken Klaus von Klitzing's measurements.

Horst was led into a cold room and asked to put on a hat (not unlike the inverted colander in the movie *Ghostbusters*), called a MEG hat. It measures tiny magnetic fields that emanate from the brain. They're made by electric impulses, the same ones that are picked up with electroencephalography. However, since the salt water in the human body doesn't affect magnetic fields, the thinking was that the MEG (magnetoencephalography) should be a better measurement of brain function than the EEG. There are more technical ways to describe it, but that's the basic idea. Everyone in the lab was eager to compare Horst's measurements with von Klitzing's. They hypothesized that Horst's "spikes" would be three times the size of von Kilitzing's (remember the "1/3 effect"?). The only result I recall was that yes, Horst still "had spikes."

The brain scan itself was given by Claudia Liaso. She and I were both interested in autism research and had a long conversation about the discovery that autistic people have excessive neuron firings in the frontal lobes, and that surgery may help to lessen typical autistic behaviors. She wondered if the finding could also be applicable to epilepsy. She had, tragically, lost her teenaged son to severe epilepsy and advised me not to take a single day with my sons for granted.

Our hotel in Helsinki was an historic edifice that had been used by both the Finns and the Nazis during WWII. The hotel was dark, gloomy, smoky, and still no doubt full of residual negative energies, so I was relieved to find a light-hearted Jimmy Stewart movie on the old television set in our room. The movie was *It's a Wonderful Life* and this was dialogue I had memorized. Hearing Stewart speaking Finnish was quite entertaining.

The next morning, we were escorted to the newly constructed Kiasma Art Museum. Helsinki itself is a city full of brightly painted buildings which offset the long, dark winter nights. The Kiasma represented a break from tradition with its arched and undulating exterior of glittering glass. An American architect named Steven Holl had won the museum's international design competition. As you enter this asymmetric edifice, its curved interior walls and sloped floor give one the impression of being cast adrift on a ship. The effect is, in fact, unsettling. The exhibits at the time were disturbing; there were videos of domestic abuse, bowls of rice lined up like soldiers, a chandelier made of bones, urban decay recreated with paintings of storefronts, and an "evil America" exhibit full of pit bulls and tattoos. The one exhibit Bob and I did enjoy was a room with a huge screen projection showing an aerial view of crashing waves. It provided an emotional cleansing I suppose. The last room we entered was full of glass vases. Each vase contained a unique scent. Walking and smelling our way across the room, gave the exhibit a playful quality that on the surface, may have been clever, but proved harmful. I smelled three scents and needed to stop. Horst smelled the contents of almost all of the vases. Holding onto the undulating railing that led downstairs, some of us began to feel dizzy and our heads throbbed. Horst began to tremble and was led into the restroom. I sat down, also feeling disoriented and ill. We returned to the hotel, where Horst recovered several hours later and managed to get through the rest of the visit.

That afternoon, I needed some fresh air and, despite the darkness, decided to venture out of the hotel to go shopping for fabrics. Finnish fabrics are, of course, uniquely designed and impeccably crafted. I was astonished by the selections of colors and textiles that were available and chose two nearly transparent white-linen square table toppers made of organdy. I also purchased two thick mohair blankets, one ranging in colors from raspberry to orange, the other designed with squares of cobalt blue and violet. The soft, rich vibrancy and understated simplicity of these textiles is truly remarkable. My purchases were also easily shipped home from the store. I imagine I had my two sons in mind as I bought the blankets. Bob and I missed the boys deeply and were growing increasingly homesick for our own rainy evening fireside chats with them.

On December 18 in Helsinki, we were treated to a lovely dinner at an elegant restaurant called Sundman's. The menu included grilled salmon with fennel salad, breast of duck with apricots, and crème brûlèe topped with cinnamon and apple sherbet. I was seated next to the famous Finnish scientist Lounismaa who was a charming dinner partner. To my left was another physicist who told me about the hazards of visiting Lapland due to mosquito overpopulation. He also explained that Christmas in Scandinavia is celebrated on the evening of the December 24. Families attend church services at 4:00am because in former days, the farmers needed to get up that early and it was the only time they could attend church. The tradition continues. I am always grateful to sit next to a physicist and not have to talk about physics. This happens rarely. Usually I am caught in a crossfire of physics lingo, the meaning of which cannot possibly be inferred by weird voice intonations or frenzied body gestures. I often sit and smile, nodding politely in a Buddhaesque manner until the banter dies down and I can insert a witty remark like: "Helsinki certainly is dark, isn't it?" When I first met Bob, I tried in vain to learn some of his physics language. One endearing phrase I learned early on was "spin flip Raman spectroscopy" which I would chant like a mantra for hours in the shower in order to sound proficient. As our sons grew older and more interested in science, each dinner conversation at home became harder for me to understand, as if the boys and their dad had conspired to shock and awe Mom with their brilliance. If it wasn't physics talk, it was computer talk with a new term surfacing each hour of the day. Sometimes an entire month would pass before I actually heard a familiar phrase. (Thus is born the term "physics widow.")

On December 19, we left Helsinki to return to Stockholm. While waiting in the airport, we spent six hours watching a BBC program that featured the ongoing impeachment hearings of President Clinton in the U.S. House of Representatives. Also reported on a split screen was the U.S. bombing of Baghdad. The BBC announcer said the bombing was ordered in response to United Nations' inspectors being turned out of Iraq. Had this response by the White House been anticipated by Sadaam Hussein? Had the confrontation been provoked? We will probably never know. It felt as if the world was beginning to unravel and our cloistered fairy-tale trip was receiving a rude reality check. Without having seen the news for several days, images of the Iraq bombings greatly upset Bob. It was a difficult scenario to take seriously at first, because it seemed to resemble the movie *Wag the Dog* a Hollywood diversion to take one's mind off the impeachment hearings.

As incomprehensible as American politics had become, we were still proud to be Americans and to return December 21 to the lofty redwoods and fragrant eucalyptus groves of northern California. Reunited with our sons, we felt our blessings multiply. In Bob's e-mail list was a long message from Aja who told us she already missed us and our extended family members. She graciously invited us to visit her in the Swedish countryside.

Our visit to Stockholm for Nobel Week was going to take a long time to assimilate. Bob and I were so grateful for the efforts of the many people who had worked so hard to make his acceptance of the Nobel Prize in Physics such a glorious part of our life's journey together. We had especially valued the time spent with fellow recipients Horst Stormer and Dan Tsui and their wives, Dominique and Linda. The Nobel experience came for us like an unexpected cosmic event, an aurora borealis of light shooting across the wintry skies of a dark December evening—brilliant and fleeting. We think back on it all with profound wonder.

AN EPILOGUE

𝕴N 2001, WE RECEIVED AN INVITATION TO RETURN TO Stockholm to celebrate the one hundredth anniversary of the Nobel Prize. Needless to say, we did not hesitate to accept and e-mailed white-tie-and-tail measurements that day! What made this particular gathering unique was that all of the living Nobel laureates had been invited to attend the celebrations. The number was approximately two hundred and fifty, with fifty-five of the laureates living in northern California at the time. We scarcely realized what a rare experience was in store for us.

Bob arrived in Stockholm on December 2, and I followed him on December 7. There were only two glitches I encountered traveling alone. The first occurred after we had all had boarded in San Francisco and my name was randomly called and I had to deplane. At the airport security were soldiers with machine guns—a precaution since the 9/11 attacks three months earlier. I deplaned nervously, as fellow passengers sighed heavily because of the delay. At the gate, I was again searched and asked to dump all the contents of my purse onto a table in front of two female guards. They looked at my pile of earrings, necklaces, bracelets, glittery hairpins, and asked: "Are you a jewelry salesman or what?"

"No," I replied calmly. "I'm going to Stockholm for the Nobel Prize ceremony."

"Oh. The what?" they replied in unison, glancing blankly at one another. "Just a minute."

They returned with two male guards who looked through my loot and must have decided none of the jewelry presented a clear and imminent danger, because I was allowed to stuff it back in my purse and re-board hearing a second chorus of heavily sighing passengers.

My next glitch occurred because I had pulled on some thick maroon-colored socks that the flight attendant had provided me with and as we landed in Copenhagen, I left them on and jammed my shoes over them. This was a huge mistake, especially since I had to jog through the airport and through security checks to a connecting flight. The flight to Stockholm had actually been held for me, so I hobbled aboard late to a third chorus of sighing passengers. By now, my heels had developed blisters the size of eyeballs and I had blood spurting down the back of both shoes. The flight attendant confiscated my socks on the way to Stockholm and tossed a dozen adhesive bandages onto my lap, which I applied liberally over both heels. When we arrived at Stockholm airport a short time later and I inserted my feet into my shoes, it felt like they were being nailed into horseshoes. Fortunately, I was greeted by an empathetic Swedish hostess who delivered me feet first to the Grand Hotel. I had brought a pair of black sling-back pumps in my suitcase that I ended up wearing comfortably for the entire week. The moral of the story is this: wear extra-large shoes when you travel and have your spouse buy you jewelry when you get there.

The format for the Nobel Centennial Week was nearly identical to 1998's except that there was a guided tour of the Nobel Centennial Exhibition at the Old Stock Exchange. This exhibit was to tour the world after the week of festivities. This was the exhibit in which we were able to hear Bob's toast on headphones as one of only five laureate speeches that were featured. That was some toast! Prior to the actual 100TH anniversary ceremony, we were able to meet Aja for lunch. We had not been assigned an attaché for this visit and it was delightful to have a relaxed visit with Aja to catch up on news.

On December 9, all of the American laureates were invited to lunch by the new U.S. ambassador to Sweden, Charles Heimbold and his wife, Monika. Lunch was held at the Hasselbacken Hotel in order to accommodate such a large number of Laureates and their spouses. The menu for this luncheon included:

Sea bass with asparagus, chanterelle and horseradish sauce
Cloudberry mousse and citrus marinated raspberries

I have hunted ever since for a cloudberry mousse recipe.

On the afternoon of December 10, 2001, we traveled once again to the illuminated City Hall, this time in a mini-bus full of Nobel laureates and their spouses. As the youngest riders on the bus, Bob and I made our way to the rear seats. In front of us sat a sea of white hair. Everyone was eerily quiet, and I remember feeling a surge of immense power—a sort of spiritual tsunami. The interior of the bus seemed to be literally pulsating with creativity, innovation, and intelligence despite the enveloping darkness of a long December evening. Memories of previous Nobel weeks were no doubt flooding back to each of the passengers as we made the measured journey once again to City Hall. After disembarking, the laureates went backstage and their spouses were seated in the center area of the auditorium. For me, sitting beside these women was an empowering feeling as well, as we endeavored to introduce ourselves to one another before the ceremonies began. Onto the stage came the royal family, including Her Royal Highness Victoria. She was dressed in a stunning ruby dress with a diamond tiara. Her mother, Queen Silvia wore a voluminous, shimmering white gown, studded with thousands of pearls. She, too, wore a diamond tiara and what appeared to be a sapphire-and-diamond necklace that radiated rainbows of light out into the audience each time she breathed in and out.

After the awards ceremony, the pageantry of the evening continued at the banquet where all the former laureates were seated at long tables perpendicular to the head table. I was seated next to Stanford colleague and 1996 Laureate Doug Osheroff, with Dominique, Horst, and 1997 Laureate Bill Phillips nearby. We had a raucous time together. I wore a pink satin gown with my recycled turquoise bolero jacket and pearls. Unfortunately, and much to my chagrin, the gown developed a plunging neckline as dinner progressed, especially each time I laughed or leaned forward. My napkin slid off my slippery satin lap constantly and Doug was kind enough to grab it in mid-flight after I explained my neckline problem. Note for next time: Velcro the gown to an undergarment, and the napkin to the gown.

During a delicious dinner, we were entertained by the Royal Opera singers, who presented arias from Verdi's operas including *Don Carlos*, *Macbeth*, *Othello*, *La Traviata*, *A Masked Ball*, *Aida*, and *Falstaff*. Following an illuminated parfait cassis meringue dessert, Bob started motioning to me as if he was getting ill. As I knew we couldn't get up from the table, I started a chain of whispers down the table to find out what was the matter. It seemed he had bitten into something that had loosened a crown on one of his teeth and he was holding the errant silver crown between his fingers. Someone at the table proclaimed: "The king has lost his crown!" I wondered if there were some caveat in the protocol rules about dental emergencies. Probably not. In any case, it didn't hold Bob back from another trip to the student Nightcap party.

The Nightcap that year was being given by the Medical Students Association of Stockholm in cooperation with the Karolinska Institute. The theme of the party was mythology and fairytales. We had taken a taxi to the party along with Dominique and Horst. When we arrived at the door, we had no tickets and of course, Horst still did not carry a passport. So we had to scramble to find some form of identification. The students did not believe they were looking at two Nobel laureates. Eventually someone was found who recognized them and we were allowed inside. We made our way downstairs first, following some dark, winding stairs. At the bottom, we discovered we were in a classroom right out of a *Harry Potter* novel. "Professor Snape" was writing incomprehensible physics problems on the blackboard while challenging guest "students" who were wedged into old wooden desks, for the answers. Bob raised his hand and was repeatedly ignored, until finally he burst out: "Call that a physics problem? You've got your signs wrong!"

"My what?" Snape snarled, looking down his nose in our direction. "Do you dare correct a senior professor, you white-haired whippersnapper? Get out, I say, before I throw you out!" he screamed, flapping his black, batlike arms.

Racing back upstairs (which was not easy to do in a plunging neckline and sling-back shoes), we transmigrated into a "Narnia" room full of sprawling tree branches and life-sized wood nymphs offering steaming elixirs. Other rooms included visits to *The Lord of the Rings*, *Alice in Wonderland*, *Sleeping Beauty*, Sherwood Forest, Nordic mythology, Greek mythology, and finally, it was on to disco dancing!

The delicious food at the Nightcap included the following:

Potato and Reindeer Tortilla with Horseradish
and Crème Fraîche
Pickled Herring with Lime and Red Onion
White Fish Roe and Marinated Herring
Pie with Swedish Cheese from Västerbotten
Traditional Swedish Egg and Herb Salad with Crisp Bread

That dinner concluded with six late-night Supper Songs that were sung with much gusto. By then, we were full of beer and undetermined elixirs (that the medical students had devised). It was impossible not to go back to the hotel without rosy cheeks and a sense of the mythical.

On December 11, after a day of packing and napping, Bob and I dressed for the royal banquet at the palace. Relaxed and looking forward to the evening, we made our way through a mobbed lobby and out onto the mini-bus transport. I had chosen to wear a sparkly lavender gown with a high neck and a full-length matching coat—no possibility for cleavage disruptions in that gown. I also wore my black sling-back shoes which had well broken-in soles, so I was reasonably confident there would be no repetition of my near-collisions with His Majesty, King Gustaf.

In the receiving line at the palace, the king recognized Bob and said: "What, you again? You're getting to be a regular here." The royal family was exceptionally friendly and despite the large number of guests (several hundred), they made everyone feel remarkably at ease throughout the evening. I cannot recall where Bob was seated for this dinner, but I was seated at the table nearest the door to the entrance. It was roughly three quarters of a mile from the head table and physically the farthest possible location from my seat in 1998. However, this was to be expected and my new vantage point afforded me the chance to admire the magnificent china and crystal collections on exhibit in the wall of cabinets behind me. That night the Royal Banquet featured:

Turbot poche
Brandade de crabe et sauce a l'estragon
Filet de chevreuil de la chasse Royale
Surprise aux trios chocolates

After dinner, the laureates enjoyed talking with one another beneath a cumulus cloud of cigar smoke and whiffs of brandy. At 10:30 or so, the royal family began to leave and we all stood to salute their departure. The evening continued on as the laureates had tremendous curiosity about one another, and no one was in a rush to leave. I was happy to prolong the visit. I wasn't certain I would ever return and felt bittersweet about our own departure. But eventually we noticed to our chagrin that we were nearly the last ones in the palace. We were two of a half-dozen hearty revelers, who logically decided there was safety in numbers, and together we all set forth to find an exit. The doors that were visible to us all appeared to be locked. An added problem, as my husband delicately described it, was "that we were all schnockered." This is a technical physics term. Feeling like eight tipsy Cinderellas at the ball, we marveled at the majestic rubiks cube of invisible doors and passageways into which servants entered, never to be seen again.

"We'll be stuck in here forever!" someone pronounced.

Finally, we located a false door and behind it a palace official who was quite amused by our predicament. Not missing a beat, he smiled and graciously said: "Let me show you out. This way, if you please." He led us past rooms full of dinner dishes and dessert goblets—all cleanly emptied of their triple-layered chocolate surprise. There was an expanse of stainless steel counters, piles of silver and buckets of soaking utensils. Rounding a corner, we passed an area set up like a mini-emergency room with a bed and medical paraphernalia. Onward through corridors and down a sort of elevator dumbwaiter, we finally were led to the level of the stairway that lead down to the entrance.

"I see where we are now!" Bob proclaimed with the confidence of a veteran Sierra mountain hiker who never carries a compass but doesn't seem to get lost.

On December 12, 2001 we reluctantly left Stockholm, after having made a number of new acquaintances among the laureates and their spouses. It is impossible to imagine any venue in the entire world that is as life-affirming as the Nobel Week. Bob and I remain humbled, and eternally grateful, to have been participants in the Nobel Prize experience.

The Laughlin Family:
Nathaniel, Todd, Anita and Bob at the Nobel Ceremony

Photo by Margaret Martin

More information can be found at
Bob Laughlin's Nobel website: **http://large.stanford.edu**
or Anita Laughlin's website: **www.anitalaughlin.net**

Appendix

Item #1: Nov. 24 Invitation to the Reception at the Swedish Ambassador's in Washington, D.C.

Items #2 A and B: Guest List for the Nov. 24 Reception at the Swedish Ambassador's in Washington, D.C.

Item #3: Nobel Banquet Schedule

Items #4 A and B: Nobel Procession

Items #5 A and B: Nobel Banquet Seating Plan

Item #6: Nobel Dinner Invitation to the Royal Palace

Item #7: Nobel Dinner Menu

The Ambassador of Sweden
and Mrs. Rolf Ekéus
request the pleasure of the company of
Professor and Mrs. Laughlin
at a Dinner
on Tuesday, November 24, 1998
at 8:00p.m.

Black tie   3900 Nebraska Ave. N.W.
R.S.V.P. (202) 467 2623/2611   Washington, D.C. 20016
To remind

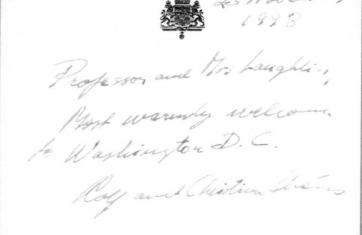

Appendix

Dinner on Tuesday November 24, 1998, at 8 pm.

In honour of the Recipients of the 1998 Nobel Awards

Professor and Mrs. Robert F. Furchgott

Professor and Mrs. Ferid Murad

Professor and Mrs. Louis J. Ignarro

Professor and Mrs. Robert B. Laughlin

Professor Horst Störmer and Mrs. Dominique Parchet

Professor and Mrs. Daniel C. Tsui

Professor and Mrs. John A. Pople

Chief Justice William Rehnquist

Justice Ruth Bader Ginsburg and Professor Martin D. Ginsburg

Mr. Joel Klein

Mrs. Katherine Graham

Mr. Robert McNamara

Dr. and Mrs. Neal Lane

Mrs. Mellanne S. Verveer

Mr. and Mrs. Sidney Blumenthal

Ambassador and Mrs. Thomas R. Pickering

Mr. and Mrs. John Newhouse

Representative and Mrs. Lee H. Hamilton

Dr. and Mrs. Harold E. Varmus

Mr. and Mrs. Michael Heyman

Mr. and Mrs. Walter L. Cutler

Mr. Stephen Trachtenberg and Mr.. Ben Trachtenberg

Professor Norman Birnbaum and Professor Leslie Griffin

Mr. and Mrs. Mark Malloch Brown

Mr. Sven Sandström

Dr. Jessica Mathews

Mr. and Mrs. Walter Pincus

Mr. and Mrs. William Safire

ITEM #2 B

Mr. David Webster and Ms. Elizabeth Drew

Mr. Marc Mathews

Mr. Christopher Hitchens and Ms. Carole Blue

Ms. Nora Boustany

Ambassador Tom Erik Vraalsen

Dr. and Dr. James Woolsey

Ambassador and Mrs. Herbert Okun

Mr. and Mrs. Gerald Nagler

Ms. Camilla Nagler

Count and Countess Peder Bonde

Ambassador and Countess Wilhelm Wachtmeister

Countess Margareta Douglas

Ambassador and Mrs. Thomas Siebert

Mrs. Meg Augustine

Mr. and Mrs. John Kunstadter

Mr. and Mrs. Charles DiBona

Mr. Frank Andersén and Ms. Anna-Maria de Mianova

Mr. Charles Reinhardt

Mr. David Furchgott and Mrs. Fetneh Fleischman

Mrs. Jane Roth and Mr. Clark Danford

Mr. and Mrs. Kurt Mälarstedt

Ms. Cecilia Uddén-Mannheimer and Mr. Otto Mannheimer

Mr. Lars Adaktuson

Ms. Karin Henriksson and Mr. Gary Yerkey

Ms. Helena Ekéus and Mr. Gustavo Zlauvinen

Mr. Oscar Ekéus and Ms. Colléen Quinn

Mr. and Mrs. Peter Tejler

Mr. and Mrs. Bo Hedberg

Ms. Nina Ersman

Mrs. Nancy Andersson

Mrs. Britt-Marie Hagelbrant Paul and Mr. Andrew Paul

Appendix

TIME SCHEDULE FOR THE 1998 BANQUET

6.45 p.m. Presentation of the Laureates and other guests of honour to Their Majesties the King and Queen of Sweden in the Prince's Gallery
*

6.30 p.m. Guests are welcomed into the Blue Hall and requested to take their seats

7.00 p.m. Guests at the table of honour enter in procession on the balustrade Fanfares. The entry procession is accompanied by an organ and trumpets

7.09 p.m. His Majesty's toast is proposed by the host of the evening's events, Professor Bengt Samuelsson, 1982 Nobel Laureate in Physiology or Medicine and Chairman of the Board of the Nobel Foundation

7.10 p.m. A toast to Alfred Nobel's memory is proposed by His Majesty the King

7.11 p.m. 4 minute photo opportunity of the heart of the table of honour

7.15 p.m. The first course is served. Music by The Stockholm Sinfonietta. Conductor Cecilia Rydinger Alin. Directly upon the serving of the first course, a 5 - minute photo opportunity at the tables of the Laureates' families and other tables (pool photographers)

8.00 p.m. The main dish is served. Music by The Stockholm Sinfonietta. Conductor Cecilia Rydinger Alin

8.55 p.m. A musical divertissement, approx. 15 minutes. Immediately after, the dessert is served. Soprano Katarina Dalayman performs.

After the dessert has been served Photo opportunities: 4 minutes of the heart of the table of honour, plus 5 minutes at the tables of Laureates' families and other tables (pool photographers)

9.50 p.m. Coffee is served.
Students from Swedish universities and colleges, bearing the standards of their student unions, pay homage to the Laureates. Fanfares before every
10.10 p.m. Laureate's speech - one speaker for each Prize category

Literature: Mr. José Saramago
Physiology or Medicine:
Chemistry: Professor John Pople
Physics:
Economics: Professor Amartya Sen

approx.
10.25 p.m. A signal is given to rise from the table

Dancing in the Golden Hall to The Kustbandet Orchestra
*

Their Majesties the King and Queen receive the Laureates.

Reindeer with King Gustaf

Procession line

I   H.M. Konungen
Mrs Joy A. Pople

II   Professor Bengt Samuelsson
H.M. Drottningen

III   H.E. Mr Jorge Sampaio
H.K.H. Hertiginnan av Halland

IV   Mr José Saramago
Talman Birgitta Dahl

V   Statsminister Göran Persson
Mrs Maria José Ritta Sampaio

VI   Professor Horst L. Störmer
Docent Karin Samuelsson

VII   Professor John A. Pople
Prinsessan Christina, Fru Magnuson

VIII H.E. Riksmarskalken, Professor Gunnar Brodin
Ms Anita P. Laughlin

IX   Professor Robert F. Furchgott
Fru Kerstin Brodin

X   Professor Robert B. Laughlin
Fru Annika Persson

XI   H.E. Mr José Socrates
Kulturminister Marita Ulvskog

XII   Professor Louis J. Ignarro
Mrs Carol Murad

XIII Professor Amartya Sen
Ms Dominique Parchet

XIV Professor Ferid Murad
Dr Sharon Williams Ignarro

XV   Utbildningsminister Thomas Östros
Mrs Emma Rothschild Sen

XVI Professor Daniel C. Tsui
Grevinnan Sonja Bernadotte af Wisborg

XVII	Sir John Vane Grevinnan Trolle-Wachtmeister
XVIII	Professor Kai Siegbahn Mrs Margaret Furchgott
XIX	Professor Klaus von Klitzing Ms Pilar Del Rio
XX	Greve Lennart Bernadotte af Wisborg Mrs Linda Tsui
XXI	Ambassadör Ketil Börde Sjuksköterskan Susanne Östros
XXII	H.E. Mr Paulo Castilho Konsulent Esther Kostöl
XXIII	H.E. Mr Roger B. Bone Mrs Brinda Dubey
XXIV	H.E. Mr Lyndon L. Olson, Jr. Mrs Kay W. Olson Professor Lars Gyllensten
XXV	H.E. Dr Michael Naumann Mrs Lena M. Bone Adjunkt Mats Ulvskog
XXVI	Generalkonsul Tord Magnuson Lady Daphne Vane
XXVII	Greve Trolle-Wachtmeister Cand. mag. Gudrun Rise Börde
XXVIII	Chefredaktör Enn Kokk Mrs Renate von Klitzing
XXIX	Förbundsleder Jan Werner Hansen Adjunkt Anna-Brita Siegbahn
XXX	H.E. Mr Sushil Dubey Mrs Maria Luisa Castilho
XXXI	H.E. Mr Klaus-Hellmuth Ackermann Mrs Adelheid Ackermann

STOCKHOLMS SLOTT

DEN 11 DECEMBER 1998

Vice ceremonimästaren Kommendör 1.gr Bertil Daggfeldt

Professor Mats Jonson	Professor Björn Roos
Fru Ann-Marie Eberson	Informationssekreterare Cecilia Wilmhardt
Professor Björn Berglund	Professor Ralf F. Pettersson
Författaren Horace Engdahl	Författaren Per Wästberg
Professor Erik Karlsson	Fru Margreta Norrby
Informationschefen Fru Agneta Nobel-Wennerström	Direktör Michael Sohlman
Professor Nils Ringertz	Överintendent Bengt Telland
Professor Anders Karlqvist	Professor Tore Frängsmyr
Professor Jan S. Nilsson	Författaren Katarina Frostenson
Direktör Barbro Fischerström	Professor Anders Flodström
Professor Claes-Robert Julander	Friherrinnan Ann Marie Ramel
Fru Gurli Lemon Bernhard	Professor Thomas Rosswall
Professor Lars Gyllensten	Fru Ylva Isaksson
Professor Sigbrit Franke	Universitetskansler Stig Hagström
Fv Stabschefen Amiral Bror Stefenson	Adjunkt Birgitta Sjöö
Fru Gunilla Storch	Förbundsleder Jan Werner Hansen
Överceremonimästaren Ambassadör Tom Tscherning	Fru Karin Stefenson
Speciallärare Solveig Allén	
H.E. Mr. Klaus-Hellmuth Ackermann, Ambassador of Germany	H.E. Mr. Sushil Dubey, Ambassador of India
Fru Anna-Brita Siegbahn	Grevinnan Sonja Bernadotte af Wisborg
H.E. Mr. L. Olson Jr., Ambassador of the United States of America	Professor Klaus von Klitzing
Ms. Jane Roth	Docent Karin Samuelsson
Mrs. Margaret Furchgott	H.E Mr. Roger Bone, Her Britannic Majesty's Ambassador
Förste livmedikus Docent Torbjörn Lundman	Lady Vane
Mrs. Luisa Castilho	Professor Amartya Sen
Greve Trolle-Wachtmeister	Cand.mag. Gudrun Börde Rise
Fru Kerstin Brodin	Statsrådet Thomas Östros
Professor Daniel C. Tsui	Ms. Dominique Parchet
Statsrådet Marita Ulvskog	Professor Louis J. Ignarro
Dr Ferid Murad	Prinsessan Christina, Fru Magnuson
Talmannen Fru Birgitta Dahl	Mr José Saramago
Professor John A. Pople	Mrs. Anita Laughlin
H.M. Drottningen	**H.M. Konungen**
Professor Horst L. Störmer	Dr. Sharon Williams Ignarro
H.K.H. Hertiginnan av Halland	H.E. Riksmarskalken Professor Gunnar Brodin
Professor Robert F. Furchgott	Ms. Pilar del Rio

Appendix

ITEM #5 B

Överhovmästarinnan Grevinnan Trolle-Wachtmeister

Professor Robert B. Laughlin

Mrs. Carol Murad

Generalkonsul Tord Magnuson

Mrs. Emma Rothschild Sen

Greve Lennart Bernadotte af Wisborg

Mrs. Lena Bone

Doktor Neal F. Lane

Mrs. Brinda Dubey

Sir John R. Vane

Mrs. Kay Olson

Direktör Marcus Storch

Statsfrun Louise Lyberg

Professor Gustaf Lindencrona

Konsulent Esther Kostol

Professor Hans Wigzell

Hovmarskalken Friherrinnan Elisabeth Palmstierna

Professor Jan-Eric Sundgren

Professor Gunnel Vallquist

Chefstandläkare Bo Wennerström

Professor emer. Carl Gustaf Bernhard

Avd. dir Anne-Marie Ringertz

Professor Torvard C. Laurent

Professor Lennart Eberson

Hovmarskalken Generalmajor Tomas Warming

Fil.lic Margit Berglund

Professor Carl-Gustav Groth

Livmedikus Peter Möller

H.M. Konungens vakth. adjutant Överstelöjtnant Ulf Gunnehed

Chefredaktör Enn Kokk

Mrs. Linda Tsui

Ambassadör Keiil Borde

Mrs. Joy C. Pople

H.E. Mr. Paulo Castilho, Ambassador of Portugal

Mrs Renate von Klitzing

Professor Kai Siegbahn

Fru Susanne Östros

Professor Bengt Samuelsson

Mrs. Adelheid Ackermann

Stabschefen Generallöjtnant Curt Sjöö

Mrs Joni Sue Lane

Friherre Stig Ramel

Fru Birgitta Tscherning

Ambassadör Krister Isaksson

Professor Boel Flodgren

Professor Erik Lönnroth

Fru Kerstin Telland

Ständige sekreteraren Professor Sture Allén

Rektor Magnificus Bo Sundqvist

Författare Östen Sjöstrand

Med. dr. Lena Karlqvist

Professor Kjell Espmark

Professor Inge Jonsson

Professor Erling Norrby

Hovdamen Fru Wendela Mattsson

Författaren Torgny Lindgren

Professor Sten Lindahl

Kammarherren Friherre August Trolle-Löwen

Förste hovmarskalken Johan Fischerström

ENTRÉE

ITEM #6

THE FIRST MARSHAL OF THE COURT IS COMMANDED BY

HIS MAJESTY THE KING

TO INVITE

Professor and Mrs. Robert B. Laughlin

to a Dinner
TO BE GIVEN BY

THEIR MAJESTIES THE KING AND QUEEN OF SWEDEN

at The Royal Palace (Eastern Wing)
on Friday, 11th of December 1998 at 7.30 p.m.

Dress: White tie
Arrival not later than 7.30 p.m. Please, bring this card.

R.S.V.P.
P.M.

DÎNER

DU 11 DÉCEMBRE 1998

Terrine de Foie gras de Canard au Vieux Madère

Fruits de Mer dans sa Nage Saffranée

Selle de Chevreuil de la Chasse Royale

Reblochon Fermier

Succès de Framboises Arctiques

★

Château Contet à Barsac 1979

Domaine Sainte-Anne 1993

Clos d'Ière 1993

LaVergne, TN USA
09 October 2009
160487LV00002B/3/P